Sustainable Textiles: Production, Processing, Manufacturing & Chemistry

Series Editor

Subramanian Senthilkannan Muthu, Chief Sustainability Officer,
Green Story Inc., Tung Chung, Hong Kong

This series aims to address all issues related to sustainability through the lifecycles of textiles from manufacturing to consumer behavior through sustainable disposal. Potential topics include but are not limited to: Environmental Footprints of Textile manufacturing; Environmental Life Cycle Assessment of Textile production; Environmental impact models of Textiles and Clothing Supply Chain; Clothing Supply Chain Sustainability; Carbon, energy and water footprints of textile products and in the clothing manufacturing chain; Functional life and reusability of textile products; Biodegradable textile products and the assessment of biodegradability; Waste management in textile industry; Pollution abatement in textile sector; Recycled textile materials and the evaluation of recycling; Consumer behavior in Sustainable Textiles; Eco-design in Clothing & Apparels; Sustainable polymers & fibers in Textiles; Sustainable waste water treatments in Textile manufacturing; Sustainable Textile Chemicals in Textile manufacturing. Innovative fibres, processes, methods and technologies for Sustainable textiles; Development of sustainable, eco-friendly textile products and processes; Environmental standards for textile industry; Modelling of environmental impacts of textile products; Green Chemistry, clean technology and their applications to textiles and clothing sector; Eco-production of Apparels, Energy and Water Efficient textiles. Sustainable Smart textiles & polymers, Sustainable Nano fibers and Textiles; Sustainable Innovations in Textile Chemistry & Manufacturing; Circular Economy, Advances in Sustainable Textiles Manufacturing; Sustainable Luxury & Craftsmanship; Zero Waste Textiles.

Subramanian Senthilkannan Muthu
Editor

Cultural Sustainability in Fashion

Springer

Editor
Subramanian Senthilkannan Muthu
Chief Sustainability Officer
Green Story Inc, Canada

Tung Chung, Hong Kong

ISSN 2662-7108 ISSN 2662-7116 (electronic)
Sustainable Textiles: Production, Processing, Manufacturing & Chemistry
ISBN 978-3-032-03206-5 ISBN 978-3-032-03207-2 (eBook)
https://doi.org/10.1007/978-3-032-03207-2

This Springer imprint is published by the registered company Springer Nature Switzerland AG
The registered company address is: Gewerbestrasse 11, 6330 Cham, Switzerland

This book is dedicated to:
The lotus feet of my beloved Lord
Pazhaniandavar
My beloved late Father
My beloved Mother
My beloved Wife Karpagam and Daughters,
Anu and Karthika
My beloved Brother
Last but not least
To everyone working in fashion space to
make the world, "SUSTAINABLE"

Contents

Cultural Sustainability in Fashion

Subramanian Senthilkannan Muthu

With today's entirely developed, globalized, and fast-paced world, our survival with our own cultural identities, practices, and traditions is becoming quite difficult, and this can be accepted by anyone in this era. As we are witnessing huge changes in our ecosystem like our cities do expand, languages are increasingly disappearing, and very importantly technology is reshaping our daily lives, our global communities are grappling with one crucial question: How do we preserve the very essence of who we are, while adapting to the fast-paced, quite advanced world around us? The answer is quite simple—cultural sustainability, which offers a reasonably clear path forward to us. It primarily challenges the assumption that progress must come at the very expense of cultural heritage and also it proposes an useful model where tradition and innovation coexist. This book is absolutely dedicated to deal with the concept of cultural sustainability and its implications in the fashion sector. This book primarily introduces the concept of cultural sustainability, explores its significance, and outlines the key elements that form its foundation, and this book revolves around the fashion sector. Cultural sustainability refers to the ongoing viability of cultural systems. It ensures that cultures do remain not just their physical representations, but their values, knowledge systems, and ways of life can endure, evolve, and be passed down in meaningful ways. While environmental and economic sustainability are often widely recognized pillars of sustainable development, cultural sustainability is often underemphasized across all industrial sectors. Yet without it, sustainable development lacks soul, identity, and social cohesion, which is the primary emphasis of this important book.

In extending the concept of cultural sustainability to fashion, one can often observe how clothing becomes more than aesthetics, it transforms into a medium for storytelling and heritage preservation. As anyone accepts, fashion is more than

S. S. Muthu (✉)
Green Story Inc, Toronto, ON, Canada

S. S. Muthu (ed.), *Cultural Sustainability in Fashion*, Sustainable Textiles: Production, Processing, Manufacturing & Chemistry,
https://doi.org/10.1007/978-3-032-03207-2_1

fabric and it is a storytelling tool. What we wear communicates our identity, status, memory, and culture. Across history and continents, clothing has played a central role in expressing who we are and where we come from mainly. Yet in today's globalized fashion industry, cultural symbols are often borrowed, commodified, or erased in the name of trend and profit. Cultural sustainability in fashion is all about preserving traditional textile arts, respecting cultural heritage, and ensuring that communities not corporations are the primary beneficiaries of their cultural contributions. Fashion can be a force for cultural erasure or for empowerment. When done sustainably, fashion honors the hands, histories, and communities behind the cloth. Cultural sustainability in fashion is not a trend, it's a necessary evolution of how we create, wear, and relate to clothing in a globally interconnected world. This important concept is disseminated in this book via five wonderful chapters written by global experts in the field.

The second chapter deals with transformation of traditional clothing to modern fashion in Tamil culture and its sustainability. As mentioned in this chapter, the Tamil culture is globally said to be one of the oldest cultures on the earth. Clothing is an important feature in the rich and diverse Tamil culture, which anyone can witness. The saree or sari and dhoti are the chief clothing types for women and men respectively in the Tamil culture. Traditionally, cotton and silk were the clothing of the poor and the rich respectively. Tamil culture developed intricate textiles by hand spinning, hand weaving, and natural dyeing with dyes derived from mostly plant sources. But the clothing of the Tamil people has transformed astonishingly to keep in line with the fashion trends of the present day. This chapter discusses this phenomenal transformation and its related aspects in detail. This chapter begins with an investigation of the transformation of Tamil attire through a cultural and historical lens, emphasizing the symbolism, craftsmanship, and resilience of traditional dress across time. It also explores the historical evolution of Tamil clothing, tracing its journey from ancient, draped garments to contemporary representations. By analyzing literary sources, temple sculptures, colonial influences, and modern reinterpretations, the study explained in this chapter reveals how clothing in Tamil culture reflects broader socio-political, religious, and aesthetic narratives. Finally, this informative chapter primarily aims to document how clothing has evolved from a functional necessity to a marker of identity and pride, while retaining continuity with its roots.

The third chapter is on an interesting topic. It talks about Innovating Tradition: Fiber-Reinforced Composites in Sustainable Fashion Design. Fiber-reinforced composites (FRCs) offer immense potential for driving sustainability in fashion by integrating advanced material science with traditional craftsmanship. Natural fiber-reinforced composites (NFRCs) represent a groundbreaking advancement in fashion design. This chapter interestingly explores the role of FRCs, particularly those reinforced with natural fibers such as hemp, flax, and bamboo, as eco-friendly alternatives to synthetic composites. By combining lightweight durability with aesthetic flexibility, these materials create opportunities for innovative fashion designs that meet both functional and cultural needs. The chapter delves into the sustainability metrics of natural fiber-reinforced composites (NFRCs), examining their environmental benefits, including reduced carbon footprints, lower energy consumption, and enhanced recyclability. It also highlights how traditional textile techniques and artisanal skills can be preserved and revitalized through their integration into FRC production, fostering cultural sustainability. Through case studies and real-world examples, the chapter shows how NFRCs can be applied in fashion apparel, accessories, and footwear, aligning durability with cultural aesthetics. This chapter delves into the potential of NFRCs to address some of the fashion industry's most pressing challenges. Through an exploration of sustainability metrics, case studies, and emerging technologies, it highlights how these materials can reduce environmental impact while fostering cultural diversity. For instance, hemp and bamboo are renewable resources with low environmental footprints, making them ideal for eco-conscious designs. Challenges such as scalability, consumer acceptance, and cost-effectiveness are critically examined, alongside emerging solutions like bio-based resins and additive manufacturing. By bridging cultural heritage and material innovation, this chapter envisions a future where sustainable fashion not only addresses environmental challenges but also serves as a platform for preserving and celebrating diverse cultural identities.

The fourth chapter talks about heritage fashion, which is perceived to be an important catalyst for sustainability. Heritage fashion refers to clothing and accessories that draw inspiration from historical styles, traditional craftsmanship, and cultural legacies, blending them with contemporary aesthetics to create timeless pieces. This approach highlights real materials, timeless designs, and a respect for cultural narratives as an alternative to the rapid cycle of fast fashion. This chapter details the intersection of sustainability and cultural heritage within the fashion industry, specifically focusing on South Asian traditions and practices. The authors of this chapter explore how heritage fashion brands constantly preserve traditional craftsmanship which includes handweaving, embroidery, natural dyeing, etc. Heritage fashion serves as a powerful catalyst for sustainability, intertwining traditional craftsmanship with ethical and eco-friendly fashion practices. This chapter highlights the significant role heritage fashion plays in preserving cultural identity while promoting environmental responsibility. By focusing on traditional techniques such as handweaving, embroidery, and natural dyeing, heritage fashion reduces reliance on synthetic materials, minimizes waste, and fosters sustainable consumption. This chapter also deals with sustainable materials like organic cotton, hemp, and plant-based dyes where their role in reducing the reliance on synthetic fabrics is increased. This chapter also presents different case studies, where fashion businesses integrate sustainability and cultural heritage preservation to gain a competitive advantage in the apparel industry which also makes a positive societal impact.

The fifth chapter is dedicated to deal around alternative textile fibers as a solution for improving sustainable fashion. Textile fibers and materials play the most significant role in obtaining a sustainable textile and fashion due to their determinative contribution to production, use, and end-of life options of textiles. Selection of the fibers to be used in a textile garment and product is of utmost importance in the design stage, affecting all details of production, processing, and end-use and the resulting sustainability and circularity of the end-product. In this chapter, the conventional fibers are discussed about their positive and negative impacts on the properties of the textile product and their advantages and disadvantages will be put forward. Insights are given about biodegradability, water consumption, energy consumption, clothing comfort, and end-of-life options. They will be compared with a wide perspective in terms of their contribution to obtaining sustainable fashion and textiles. Fiber alternatives to conventional fibers will be evaluated. Recycled cotton and recycled polyester are to be discussed besides sustainable fibers including fibers obtained from fruits (banana, orange, pineapple, coconut, coffee, mango), fibers obtained from vegetables (avocado, artichoke, etc.), and fibers such as bio-based polyester obtained from renewable sources as alternative to synthetic fibers obtained from nonrenewable sources. Besides all these fibers, a big emphasis will be given to more sustainable man-made cellulosic fibers. Natural fibers including kapok, jute, and hemp will also be discussed with their features and their possible place in the fashion industry. In light of these developments, it is essential to critically examine the environmental impacts associated with conventional textile fibers and evaluate the potential of bio-based and recycled alternatives as more sustainable solutions. Therefore, in this chapter, the environmental impacts of both natural and synthetic textile fibers are discussed in detail by highlighting their functional features and the challenges starting from the manufacturing and end-of-life stages. In response to the growing need for sustainable alternatives, this chapter further explores a range of innovative fiber solutions including bio-based synthetic fibers such as Sorona® (bio-poly(trimethylene terephthalate)), polylactic acid (PLA), and bio-based polyethylene terephthalate (bio-PET); recycled fibers like recycled polyester (rPET) and recycled cotton; and a variety of sustainable natural-origin fibers such as corn fiber, Gallus-Gallus feather fibers, lotus fiber, pineapple leaf fiber, orange fiber, and lyocell. Each of these alternative sustainable textile fibers is examined with a focus on their environmental advantages, material properties, and capacity to contribute to a more circular and responsible textile economy. Supported by ongoing scientific research and industry initiatives, these innovations play a pivotal role in driving the transition toward an environmentally conscious and circular fashion industry.

The sixth chapter is about traditional textile based contemporary fashion and how this can be perceived as a roadmap to cultural sustainability. Cultural sustainability is a crucial pillar of sustainable goals and principles impacting the social, economic, and environmental aspects and primarily focusing on rectifying the biases of the past and reforming the current system for diverse, sustainable discourse in fashion. Cultural sustainability pertains to the access, usage, and adaptation of cultural heritage ensuing that communities who owe the flagship of indigenous and pristine craft and artforms are not deprived of their due rights and do not face any exploitation in the process. It thus becomes crucial to shift the narrative focusing on the skill and labor that goes into fine craftsmanship and continued craft innovation. Cultural sustainability is an umbrella term and can be considered as fourth pillar of sustainable goals and principles with emphasis on:

- Cultural and artistic practices

- Knowledge dissemination and cultural heritage conservation
- Role of cultural traits and actions determining the sustainable livelihood of artisans

In the Indian context, each state has special identity for its unique traditional and hand-crafted textiles diversified talents, interests, and inspiration. Most of the traditional techniques are performed on handloom fabrics; however, a range of raw materials can be experimented with to bring out the magnificence of traditional art forms on textiles. Apparel and accessory designers take inspiration from the lively, mythological, and colorful motifs of these beautiful art forms and have been extensively using these motifs for design and development of a gamut of high-end artifacts, apparels, home textiles, and accessories using a range of textile raw materials apart from conventionally used handlooms fabrics. This important chapter discusses the roadmap followed by designers to work in association with artisans on traditional textiles and reviving the pristine forms via contemporary and modernized makeover of attire. Accordingly, the preferential traditional techniques which entice consumers and flaunt of comfort, dexterity stretch ability were identified in the study. Furthermore, the methodologies adopted by various fashion brands and designers for endorsing the traditional textiles and artisans' masterpieces on a global platform are also highlighted in this chapter.

I take this opportunity to thank all the contributors for their earnest efforts to bring out this book successfully. I am sure readers of this book will find it very useful.

With best wishes,

Dr. Subramanian Senthilkannan Muthu

Transformation of Traditional Clothing to Modern Fashion in Tamil Culture and Its Sustainability

K. Amutha and S. Thangamani

1 Introduction

Clothing is considered as a basic necessity for human life. Digging deep into the origin of clothing did not yield any evidence-based information. But an apparent fact is that humans have been using clothing in one form or the other, for example early humans wore leaves and grasses as clothing and during the stone age when humans started hunting animals their skin were used as clothing especially to protect from adverse weather conditions. Further, archeological evidence show that clothing emerged as a vital adaptation for human life.

From Darwin's theory to the current science and technology, human's life has evolved exponentially in all aspects of life. Clothing holds a special role in human life. Across the world, there are a wide array of human civilizations that differ based on continent or country or region. Clothing is intimately linked to the culture of a certain civilization or region or community. Clothing could be claimed as a cultural identity.

Clothing often serves as a visible expression of cultural identity, as clothing choices can reflect a person's ethnic background, traditions, and heritage. Specific clothing styles, patterns, and colors can act as powerful cultural signifiers, conveying a wealth of information about a person's identity and cultural affiliations. Clothing choices can reflect personal tastes, beliefs, values, and cultural backgrounds, making fashion a powerful tool for expressing one's identity. People often

K. Amutha (✉)
Department of Textiles and Apparel Design, Bharathiar University, Coimbatore, Tamil Nadu, India
e-mail: amuthatad@buc.edu.in

S. Thangamani
Department of Tamil, Bharathiar University, Coimbatore, Tamil Nadu, India

S. S. Muthu (ed.), *Cultural Sustainability in Fashion*, Sustainable Textiles: Production, Processing, Manufacturing & Chemistry,
https://doi.org/10.1007/978-3-032-03207-2_2

use clothing as a canvas to present the image they want to portray, signaling their uniqueness, values, and interests to the outside world [1].

Every country or region has different characteristics as an identity that is manifested in the form of traditional or distinctive clothing, customs, folk songs, and language. Clothing has a variety of explicit and implicit meanings. The clothes worn by someone have hidden meanings and messages, even clothes no longer only function as protectors but also to show their social status in society. Clothing was originally designed as a way to protect ourselves from cold and other elements [2].

Among the historically old and rich cultures across the globe, the Tamil culture occupies a significant position. Tamil culture, rooted in the Dravidian language family, has a rich history tracing back to the ancient inhabitants of the Indian subcontinent, with the Tamil language having a written history dating back over 2000 years to the Sangam period (300 BCE to 300 CE).

This chapter investigates the transformation of Tamil attire through a cultural and historical lens, emphasizing the symbolism, craftsmanship, and resilience of traditional dress across time. It also explores the historical evolution of Tamil clothing, tracing its journey from ancient draped garments to contemporary representations. By analyzing literary sources, temple sculptures, colonial influences, and modern reinterpretations, this study reveals how clothing in Tamil culture reflects broader socio-political, religious, and aesthetic narratives. It aims to document how clothing has evolved from a functional necessity to a marker of identity and pride, while retaining continuity with its roots.

2 Tamil Culture

Tamil culture, with its millennia-old legacy, is a tapestry woven with language, art, music, architecture, and clothing. Among these, traditional attire plays a vital role in conveying identity, social status, gender roles, and regional affiliation. The way Tamil people have dressed across centuries is not only a reflection of climatic and material conditions but also an embodiment of aesthetic values and philosophical beliefs.

From the early Sangam period to the present day, Tamil clothing has seen shifts influenced by political changes, religious movements, colonial encounters, and globalization. The fabrics, draping styles, weaving techniques, and decorative elements carry within them layers of meaning—often signifying a dialogue between tradition and transformation.

Tamil people of India are well known for their rich culture that is well understood from the Tamil literature, clothes, art and craft forms, temples, sculptures, and so on. The Tamil women gave much importance to clothing and jewelry. The modern fashion is inclusive of accessories like jewels, handbags, belts, watches, gloves, hats, scarves, etc. Similarly, in Tamil culture accessory (*anikalan* in Tamil) like jewels or ornaments always accompanied the clothing.

Though clothing is a basic necessity for human life, it exemplifies as a symbol of a specific culture. One can take food of his or her choice but dressing styles depend on other's preferences. A Tamil proverb says that "a human is incomplete without clothes" (if translated literally from Tamil: *A naked man is half a man*). The situations that signify wearing clothes are (1) defend one's honor; (2) beautify oneself; (3) protect oneself from different climatic conditions; (4) expression of artistic sensibility; (5) showing off one's improvement. The changes in politics, economy, time, and technology have direct impact on clothes and dressing styles of people. For example, the Thoda tribes of Nilgiris, India, drape clothes all over the body from neck to ankles while the tribes of Andaman Islands, India, just wear a fiber strip around the waist and leave the entire body naked expressing two extremities [3].

In Tamil literature *udukai*, *uduku*, *kupayam*, *meipai*, *pattudai*, *kalingam*, *kachu*, *danai*, *padam* were the words used to refer to clothing during the Sangam period that dates from 300 BCE to 300 CE [4].

3 Ancient Tamil Clothing (Sangam Period: 300 BCE–300 CE)

The earliest references to Tamil clothing are found in Sangam literature, such as *Purananuru*, *Akananuru*, and *Silappatikaram*. These texts describe a society that predominantly wore unstitched garments. Men typically donned a *veshti* (a long piece of cloth wrapped around the waist) and often went bare-chested or wore a thin upper cloth (*sattai*). Women wore a *pudavai* (early form of saree), which was draped in various ways without a blouse. Clothing was not just functional but expressive of one's mood, marital status, and social position.

உண்பது நாழி, உடுப்பவை இரண்டே (புறம்: 189)

The *Purananuru* verse [5] illustrates that both men and women in ancient Tamil Nadu wore a two-piece garment. While one piece was draped around the waist, the upper body was covered loosely, often leaving much of it exposed, reflecting the dressing norms of the time when partial or full exposure of the upper body was culturally unremarkable. This style of unstitched clothing, worn loosely, was gender-neutral, highlighting the simplicity and practicality of the attire. The garment's draping style, however, was not uniform; it varied based on factors such as caste and the local climate.

Natural fibers like cotton were extensively used, and evidence of silk usage is also noted in elite circles. The clothing was often adorned with intricate borders and complemented by elaborate jewelry. Both men and women used aromatic oils and flowers to complete their attire, reinforcing a strong connection between appearance and cultural identity.

Women wore a *pudavai* (early form of saree), which was draped in various ways without a blouse. Clothing was not just functional but expressive of one's mood, marital status, and social position. Women in ancient Tamil society primarily wore clothes woven from naturally biodegradable materials such as cotton or silk and embellished with flowers, leaves, and shoots of *neithal*, *aambal*, *nochi*, and *kuvalai* [6].

Historian Nivedita Louis notes that these *thazhaiyaadai* were commonly worn during festivals, sports, and even while bathing in water bodies, reflecting both their cultural significance and practical use in daily life. Natural fibers like cotton were extensively used, and evidence of silk usage is also noted in elite circles. The clothing was often adorned with intricate borders and complemented by elaborate jewelry. Both men and women used aromatic oils and flowers to complete their attire, reinforcing a strong connection between appearance and cultural identity.

In Tamil, *naar* means bast fiber. Long fibers extracted from the snake plant, *Sansevieria zeylanica* (*maral naar* in Tamil), has been used to tie strands. These fibers were extracted from the plant by water retting method. These fibers were starched and twisted and used as strings in a musical instrument called *yaazh* in Tamil [7].

The flowers and fibers of a fragrant climber plant called Indian lavanga, *Luvunga scandens* (*naravam* or *naraikodi* in Tamil) has been used by Tamil people as mentioned in *Kurunjipaatu*. The fibers extracted from this plant (called *marai naar* in Tamil) has been used to tie flowers into strands (vaiyan.blogspot.com).

Sangam literature portrays that barks of trees were worn as clothes. Leaves and flowers were made into strands using ***maral naar*** or ***marai naar*** and worn by women. These fibers have been used to weave baskets, nets, etc. Later, the technology developed into hand spinning and weaving of cotton fibers. Cotton weaving has been mentioned as a profession in Sangam literature. *Natrinai* (in Tamil) mentions that women were involved in harvesting cotton, removing the seeds from the fiber, beating the cotton with sticks to remove impurities, and then spinning it into yarn. It is very interesting that these women were literally called ***the Cotton Women*** who worked day and night with this. When they work at night, they lit small earthen lamps that glow with cotton wick and oil as mentioned in *Puranaanuru* [8]. Hence, no wonder the modern Tamil women work in day and night shifts in the textile and apparel industries.

Cotton clothes were imported to Tamil Nadu from Kalinga Nadu (now it is Odisha, a state in India) and hence they are called ***kalingam*** as mentioned in *Purananuru*. It could be understood that Tamil Nadu was considered as a country and hence the term "import" has been used for goods bought from other states of India. Silk was imported from China. Soft and delicate fabrics comparable to snake skin and the smoky inner layer of bamboo were woven by skilled weavers. Red and blue colors were used to dye the clothes. Kings and the rich people wore beautifully embellished silk clothes and the kings donated silk clothes to poets who wrote poems on praise of the king. The clothes worn by the rich were usually made by the slaves and the worn clothes were washed in the river and dried by another community of low-class people. The washed clothes were infused with fragrance (called *akirpugai* in Tamil) and then folded and kept in boxes until next use [8].

The manufactured cotton and silk clothes and the blankets were all infused with fragrance before they went for sale. A wood (agarwood) called *akir* in Tamil was burnt and the smoke has been used as fragrance for clothes. The same has been used to infuse fragrance in the long hair of women [9].

4 Clothing Tradition of Tamil People

According to a Tamil literature called *Perungathai*, the economy of a country was decided by the possessions (clothes, jewels, fragrant flowers) of women and their happiness. Also, in Tamil literature and epics, women were praised for their beauty in which the clothes and accessories have a major role. *Purananuru*, a collection of 400 heroic poems, states that clothes add beauty to the natural beauty of women and handloom was an important occupation of Tamil people. Cotton, silk and other plant fibers were used to make clothes. Very soft, white clothes were made which were as fine as the "smoke like texture seen during ocean evaporation". Women's clothes were enriched with workmanship in the form of flower embroideries. New clothes were enjoyed by the people. Colors like blue, emerald, green, and red were used to dye the clothes. Very fine and supple cloth comparable to snakeskin and smoke were woven by skilled weavers. Starch (called as *kaadi* in Tamil) was used to stiffen the fabric after laundering. Aromatic powders (*javvathu*, *athar*, etc.) were used to add good fragrance to the clothes. Then the clothes are folded and stored in box/case until further use [10].

Women draped a long piece of cloth called ***mundranai*** which draped down from the neck. A piece of cloth used to cover the breast is called ***kaccu*** or ***kachchu*** that was dyed colorfully and embellished with flower embroidery and colorful stones. Inner garment worn up to the knee is called ***vattudai***. Women wore premium clothes called ***thazhaiudai*** (literally means clothes with leaves) made of silk and cotton embellished with aromatic leaves and flowers. The types of clothes and the way to wear them are explained in *Perungathai* similar to Sangam literature [10].

The modern fashion is inclusive of accessories along with clothing. Similarly, historical Tamil clothing gave importance to accessories like jewels and footwear. Jewels were made of gold and gem stones. Nine types of gem stones (diamond, ruby, emerald, pearl, red coral, yellow sapphire, blue sapphire, cat's eye, and hessonite) popularly called as ***navaratnam*** (*nava* means nine and *ratnam* means gems) were used in jewels. Jewels were worn from head to toe. The intricate jewels embellished with red coral and beads which were worn on the waist were called ***mekalai*** also known as ***padukaal***. Another jewel called ***paandil*** made of gold and red coral was worn around the waist. Women made strands of gold coins and worn as jewel. Jewels were made in the shape of peepal leaf, which were worn to cover the private part of women. Jewels were important for a woman during her wedding ceremony and they wore many jewels.

The jewel worn on neck which extends till the chest is called ***otraivadam*** or ***aekavalli***. It is usually made with gold and pearl and sometimes with other gem stones. The jewels worn on the neck has various names like *kalikai*, *muthucharam*, *muthuvadam*, *ponnaan*, *sangili*, *ponnaram*, and *thali*. The jewels worn on wrist or forearm are called *valai*, *kadakam*, and *soodakam* that are made with gold, red coral, and tiny conch shells. The jewels worn on the fingers is called *mothiram* or *anguli*. The jewels worn on ears are called *kuzhai*, *kadippu*, *poondhodu*, etc. The jewels worn on head are named as *netrichutti*, *koondhal pattam*, *ponnarimaalai*, and *soolamani*.

The footwear made of gold is called ***gingin***. A special type of jewel worn on the anklets is called ***silambu***, which is embellished with emerald or pearl. Other types of jewels worn on legs are called ***padagam***, ***pariyagam***, ***noopuram***, and ***sathangai*** [10].

Children's clothing was not well described in Tamil literature. But in *Purananuru* it is mentioned that to send her son to war a mother tied up clothes on her son's waist as he did not know to tie properly. In the epic *Manimekalai*, the innocence and the beauty of insufficiently dressed children have been expressed poetically [3].

The men's clothing is called by different names in different literatures, which means the clothing trends varied periodically. Men's clothing was categorized based on their role in the society: king, warrior, sage or monk, and poor are some examples. During the Sangam period, kings wore long clothing that extended up to the floor, which was perceived as his pride. Such long clothing is held by the king on his left hand. It is evident from the Sangam literature that the warriors wore upper clothes with armor as a protective shield during the war. In Tamil, ***kavasam*** means a protective clothing that covers the entire body. Sages and monks wore the hides of deer and tiger which is known from *Thirumurukatrupadai*. Then in the later stage, monks mostly wore clothes in saffron color (called as ***kaavi*** in Tamil) while some monks wore white clothes. The general public has been classified as high- and low-class people based on their profession and economic status. The clothing of the high-class people is called *kalingam*, *kazhagam*, *koorai*, *meipai*, *porvai*, and *uthariyam* in literatures like *Silappathikaram* and *Perungathai*. The clothing of the low-class people is called *maranaar*, *udukkain*, *maasunudukai*, *karunthuvaraadai*, *sennthuvaraadai*, and *neelakachaadai* in various literatures. The clothing of women is also represented by various names. The women draped unstitched cloth around their body. Clothing of queens are called *poonthukil*, *kalingam*, *kodi nunthukil*, etc. Similar to the king, the queen also dressed up with clothes that extended up to the floor. The clothes of low-class women are called *thukil*, *kalingam*, *kachai*, *thazhaiaadai*, and *vambu*. These clothes were similar to the ones worn by the high-class women [3].

5 Temple Sculptures and Iconography (Sixth to Thirteenth Century CE)

With the rise of temple architecture during the Pallava, Chola, and Pandya dynasties, clothing styles were immortalized in stone. The sculptures of deities, dancers, and devotees on temple walls provide a visual archive of Tamil clothing traditions during this period. These figures often depict highly stylized garments with elaborate pleats, waistbands, and ornate jewelry.

Male figures are shown wearing *veshtis* with decorative belts, sometimes folded into multiple layers to allow ease of movement, especially for warriors and dancers. Female figures are adorned in intricately pleated sarees with front and side knots, and upper garments appear more frequently, especially in divine or royal depictions. These styles emphasize both grace and modesty, rooted in aesthetic ideals and spiritual symbolism.

Headgear, elaborate hairstyles, and various forms of *angavastram* (upper cloth) and *uttariyam* (shawl-like drapes) also become common in these representations, signifying a layered and ceremonial approach to dressing. The prominence of ornamentation in these sculptures underscores the cultural emphasis on adornment and visual richness [11].

6 Medieval to Pre-Colonial Era (Thirteenth to Eighteenth Century CE)

During the medieval period, Tamil clothing began to absorb influences from external trade, regional interactions, and the gradual spread of Islam. This era witnessed the growing popularity of finely woven textiles and dyeing techniques, including the use of natural indigo, turmeric, and madder. Trade with Southeast Asia, the Middle East, and inland kingdoms brought in new materials and motifs.

The integration of Muslim sartorial elements is visible, especially in urban centers. Men began wearing turbans and tunics over their *veshtis*, while women in some regions adopted half-sarees or wore upper garments more frequently. Tailoring became more prominent, leading to stitched garments like the choli (blouse) for women and *angarakha* or *jubba* for men.

Weaving centers such as Kanchipuram, Madurai, and Thanjavur flourished during this time. The famed Kanchipuram silk sarees, with their temple borders and mythological motifs, gained prominence not just as attire but as heirlooms and ceremonial symbols. Traditional patterns often carried philosophical meanings, representing fertility, protection, and cosmic order.

Despite these adaptations, the essence of draped garments persisted, illustrating a cultural preference for continuity even amidst change. Clothing remained a means of upholding tradition while also adapting to new influences, showing a unique cultural resilience.

7 Colonial Encounter and Impact (Eighteenth to Twentieth Century CE)

The arrival of European colonial powers, particularly the British, brought profound changes to the social, economic, and cultural fabric of Tamil society, including its modes of dress. Colonial influence led to the proliferation of Western tailoring techniques and the introduction of new fabrics and dress codes that began to challenge traditional clothing norms.

One of the most significant events during this period was the "Upper Cloth Controversy" in the nineteenth century, where Dalit women, especially from the Nadar community in southern Tamil Nadu, fought for their right to wear an upper garment in public. This movement highlighted the intersection of caste, gender, and clothing, showing how attire became a site of resistance and self-assertion.

Western-style blouses, trousers, shirts, and coats became common among the urban elite and professionals, especially those involved in colonial administration or missionary education. However, traditional garments like the *saree* and *veshti* continued to be worn with regional adaptations, sometimes blending Western elements like stitched blouses or jackets.

The colonial period also saw the commercialization of the textile industry, which had a mixed impact. While it disrupted some indigenous handloom traditions due to the influx of mill-made cloth, it also laid the foundation for a more organized textile sector. Tamil handloom weavers adapted by creating new designs and motifs that appealed to both traditional and modern sensibilities.

Clothing during the colonial era became a visual marker of social mobility, education, and political alignment. It also became a means of expressing nationalist sentiments—leaders like Mahatma Gandhi emphasized wearing *khadi* as a symbol of resistance to British rule, influencing Tamil political and cultural consciousness as well.

8 Post-Independence and Contemporary Trends (1947–Present)

The post-independence era marked a resurgence of interest in indigenous textile traditions, driven by nationalist pride and cultural revivalism. Handloom sarees, such as those from Kanchipuram, Chettinad, and Coimbatore, became emblematic of Tamil identity. Government initiatives supported traditional weaving communities, helping preserve these arts amid industrial competition.

With urbanization and globalization, Tamil clothing underwent further transformation. The saree continued to be a staple, especially during festivals, weddings, and religious ceremonies, while daily wear for women expanded to include churidar-kurta, salwar kameez, and Western attire. For men, formal and casual dressing now includes trousers, shirts, and suits, though *veshti* remains significant during cultural and religious events.

Contemporary designers are reimagining Tamil textiles and motifs, blending traditional aesthetics with modern cuts. Films, television, and social media have also played a role in shaping fashion choices, making traditional clothing both accessible and aspirational for younger generations.

At the same time, clothing remains deeply symbolic in Tamil people's life—from the ritual purity of temple wear to the performative attire of Bharatanatyam dancers. Fashion in Tamil Nadu today represents a continuum, where heritage and innovation coexist, reflecting the dynamic spirit of Tamil identity.

9 GI of Tamil Clothing

A GI or Geographical Indication is a name or a sign given to certain products that relate to a specific geographical location or origins like a region, town, or country. Using Geographical Indications may be regarded as a certification that the particular product is produced as per traditional methods, has certain specific qualities, or has a particular reputation because of its geographical origin.

Geographical indications are typically used for agricultural products, handicrafts, wine and spirit drinks, foodstuffs, and industrial products. GI Tag ensures that none other than those registered as authorized users are allowed to use the popular product name. In order to function as a GI, a sign must identify a product as originating in a given place.

GI textile	Images	Image source
Arani silk saree	 	https://cooptex.gov.in/index.php?route=product/category&path=103
Bhavani Jamakkalam	 	https://www.cooptex.gov.in/index.php?route=product/category&path=213
Kanchipuram silk saree	 	https://kanchivml.com/weave/kanchipuram-sarees?p=2&srsltid=AfmBOordulOMhWo7qfFSIvUTgLD7ZPS4mh91RHtocEE1jzfo81KwSi7n
Madurai Sungudi saree	 	https://www.cooptex.gov.in/index.php?route=product/category&path=70_101

GI textile	Images	Image source
Kovai Kora cotton saree	 	https://www.cooptex.gov.in/index.php?route=product/category&path=143_127
Salem silk known as Salem Venpattu	 	https://www.cooptex.gov.in/index.php?route=product/category&path=61_146
Thirubuvanam silk sarees	 	https://cooptex.gov.in/index.php?route=product/category&path=121
Toda embroidery		https://www.thehindu.com/society/history-and-culture/unesco-launches-list-documenting-50-iconic-indian-heritage-textiles/article65950852.ece

10 Conclusion

The journey of Tamil clothing from the unstitched drapes of the Sangam period to the curated contemporary wardrobes of today reflects a broader cultural narrative of continuity and change. Each era introduced new materials, motifs, and meanings to the tradition of Tamil dress, yet the core ethos—of elegance, symbolism, and identity—has endured. Traditional attire continues to inspire, whether as a symbol of resistance, celebration, or pride.

This exploration underscores the importance of clothing as a lens to understand history and identity. In Tamil Nadu, garments are not merely stitched fabrics but vibrant expressions of belief, systems, socio-political dynamics, and cultural resilience. Preserving and promoting these traditions in an age of fast fashion remains both a challenge and a necessity.

References

1. Das, A. (2023). *Fashion and identity: An exploration of clothing's role in shaping personal and cultural identities*. https://www.iiad.edu.in/the-circle/fashion-and-identity-an-exploration-of-clothings-role-in-shaping-personal-and-cultural-identities/
2. Tajuddin, F. N. (2018). Cultural and social identity in clothing matters "different cultures, different meanings". *European Journal of Behavioral Sciences, 1*(4), 21–25.
3. Senthilkumar, N. P. (n.d.). Clothing and dressing styles of the Sangam people (In Tamil). *Muthukamalam Online International Tamil Journal*. https://www.muthukamalam.com/essay/literature/p98.html
4. Sharmila, A. (2022). Retro clothing in Tamil Sangha literature. *International Journal for Research in Applied Science & Engineering Technology, 10*(XI), 933–936.

5. Rajeswari, C. (2022). *Puranaanuuru* (In English). Poem 189. Chandrothayam Publications.
6. Herbert, V. (2015). *Natrinai* (In English). Poem 54. Digital Maxim LLC.
7. Sundaram, V. P. K. (1985). *Tamilisai valam* (In Tamil). Madurai Kamaraj University.
8. Perumalsamy, V. (2010). *Clothes during the Sangam period* (In Tamil). https://www.keetru.com/index.php/2018-01-12-05-57-50/2014-03-08-04-39-26/2014-03-14-11-17-85/2012-2010-01-17-06-48-00
9. Varatharasan, V. (1986). *The makeup artistry of Tamils* (In Tamil). Tanjore Tamil University. https://www.tamildigitallibrary.in/admin/assets/book/TVA_BOK_0007465_தமிழரின்_ஒப்பனைக்_கலைத்திறன்.pdf
10. Devi Meenakshi, M. R. (2020). Women's attire and costumes in the Perungathai. *International Research Journal of Tamil, 2*(3), 106–116.
11. Mahalakshmi, R. (2007). Clothing the God: Temple Attire in Tamil Nadu. *Indian Economic & Social History Review*.

Innovating Tradition: Fiber-Reinforced Composites in Sustainable Fashion Design

Malik Abdul Karim

1 Introduction

Natural fiber-reinforced composites (NFRCs) represent a groundbreaking advancement in fashion design. These materials are not only lightweight and durable but also highly adaptable in terms of aesthetics, enabling designers to experiment with creative forms and textures [1]. Beyond their functional benefits, NFRCs carry profound cultural significance. By incorporating fibers that have been used for centuries in traditional textiles, these composites allow designers to craft garments and accessories that tell stories of heritage and innovation. Each piece becomes a celebration of cultural identity, blending time-honored practices with contemporary techniques [2].

By merging scientific progress with traditional artistry, NFRCs pave the way for a more sustainable and inclusive future [3]. They envision a world where fashion serves as both a solution to environmental crises and a medium for celebrating diverse cultural identities. In doing so, they exemplify the harmonious intersection of innovation and tradition, proving that style and sustainability can coexist seamlessly in the modern age [4].

This chapter delves into the potential of NFRCs to address some of the fashion industry's most pressing challenges. Through an exploration of sustainability metrics, case studies, and emerging technologies, it highlights how these materials can reduce environmental impact while fostering cultural diversity. For instance, hemp and bamboo are renewable resources with low environmental footprints, making

M. A. Karim (✉)
Division of Integrative Systems and Design, The Hong Kong University of Science and Technology, Sai Kung District, New Territories, Hong Kong
e-mail: malikkarim@ust.hk

S. S. Muthu (ed.), *Cultural Sustainability in Fashion*, Sustainable Textiles: Production, Processing, Manufacturing & Chemistry,
https://doi.org/10.1007/978-3-032-03207-2_3

them ideal for eco-conscious designs. Moreover, their integration into fashion production supports local economies and preserves artisanal skills.

2 Understanding Fiber-Reinforced Composites

Fiber-reinforced composites (FRCs) are engineered materials composed of two main components: reinforcing fibers and a matrix material [5]. The fibers, which provide tensile strength and stiffness, are embedded within the matrix, which acts as a binder to hold them together, transfer loads, and shield the fibers from environmental damage [6]. Historically, synthetic fibers like carbon and glass have been the go-to choice for FRCs due to their superior mechanical properties, such as high strength and rigidity. These materials have been widely used in industries ranging from aerospace to automotive manufacturing [7].

However, the environmental impact of synthetic fibers has raised significant concerns [8]. Their production is energy-intensive, contributing to greenhouse gas emissions, and they are non-biodegradable, leading to long-term waste management challenges. Additionally, these fibers rely heavily on fossil fuels, further exacerbating their ecological footprint [9]. As sustainability becomes a global priority, there is growing interest in natural alternatives, such as hemp, flax, and bamboo [10]. These natural fibers offer comparable mechanical performance while being renewable, biodegradable, and less resource-intensive to produce. By shifting toward natural fiber-reinforced composites (NFRCs), industries can reduce their environmental impact and promote eco-friendly innovation, aligning material science with the urgent need for sustainable practices [11–13].

2.1 Natural Fibers: A Sustainable Alternative

Natural fibers, sourced from plants such as hemp, flax, jute, and bamboo, present a compelling alternative to synthetic fibers, offering numerous environmental and practical advantages. One of the most significant benefits of natural fibers is their biodegradability [14]. Unlike synthetic fibers, which persist in landfills for centuries due to their non-biodegradable nature, natural fibers decompose naturally at the end of their lifecycle. This decomposition process reduces waste accumulation and minimizes the long-term environmental burden, making natural fibers a more sustainable choice for industries seeking to lower their ecological impact. Another key advantage is their renewability [15]. Natural fibers are derived from plants that can be cultivated and harvested annually, ensuring a consistent and replenishable supply. In contrast, synthetic fibers are typically petroleum-based, relying on finite fossil fuel resources that contribute to resource depletion and environmental degradation. By utilizing renewable plant-based materials, industries can reduce their dependence on nonrenewable resources and promote more sustainable

Table 1 Illustrates the comparative environmental impacts of synthetic and natural fibers

Aspect	Synthetic fibers	Natural fibers
Source	Petroleum-based (e.g., polyester, nylon)	Plant/animal-based (e.g., cotton, wool)
Production energy	High (energy-intensive manufacturing)	Low to moderate (depends on farming)
Water usage	Low	High (e.g., cotton requires)
Biodegradability	Non-biodegradable (long-lasting waste)	Biodegradable (eco-friendly disposal)
Durability	High (resistant to wear and tear)	Moderate (depends on fiber type)
Environmental impact	High (microplastic pollution, CO2 emissions)	Low to moderate (pesticides, land use)

production practices [16]. Additionally, natural fibers boast a significantly lower carbon footprint compared to their synthetic counterparts. The cultivation and processing of these fibers generally require less energy, resulting in reduced greenhouse gas emissions during production. For instance, plants like hemp and bamboo grow rapidly with minimal inputs, further enhancing their sustainability credentials. Collectively, these attributes—biodegradability, renewability, and a low carbon footprint—make natural fibers an environmentally friendly alternative to synthetic materials. Their adoption not only addresses critical environmental challenges but also aligns with global efforts to transition toward sustainable systems that prioritize ecological balance and resource efficiency [17] (Table 1).

2.2 Applications in Fashion

In the realm of fashion, natural fiber-reinforced composites (NFRCs) have emerged as a groundbreaking innovation, offering designers an expansive canvas to explore new dimensions in garment and accessory creation [18]. These materials are revolutionizing the industry by enabling the production of both structured garments and flexible accessories, seamlessly blending functionality with aesthetics. The versatility of NFRCs lies in their ability to be molded into various shapes and forms while retaining strength and flexibility, qualities that traditional textiles often lack [19]. This unique characteristic allows designers to push creative boundaries, experimenting with textures, silhouettes, and intricate detailing that were once deemed unattainable. For instance, bamboo-reinforced composites serve as a prime example of how NFRCs can elevate fashion design. Bamboo fibers, known for their natural sheen and softness, can mimic the luxurious drape and elegance of silk but with enhanced durability and sustainability. Such innovations not only cater to evolving consumer demands for eco-friendly options but also address practical concerns like longevity and performance [20].

The applications of NFRCs extend far beyond mimicking existing materials; they open up entirely new possibilities in fashion design. Designers can now create avant-garde pieces that combine organic textures with futuristic elements, such as sculptural dresses or architectural outerwear that defy conventional norms. These composites allow for precise control over material properties, enabling the creation of fabrics that are lightweight yet robust, breathable yet insulating. For example, flax or hemp-based composites can be engineered to provide excellent tensile strength while maintaining a soft hand feel, making them ideal for tailored suits or structured jackets. Similarly, jute-reinforced materials can be used to craft rugged yet stylish footwear or handbags that appeal to environmentally conscious consumers [21].

Moreover, NFRCs enable the integration of advanced features such as moisture-wicking properties, thermal insulation, or even embedded smart technology, making them suitable for high-performance sportswear or avant-garde couture. Imagine activewear that adapts to body temperature or a gown embedded with LED lights powered by sustainable energy sources—all made possible through NFRCs. Beyond garments, these composites are ideal for crafting accessories like handbags, shoes, and jewelry, where structural integrity meets lightweight comfort. By leveraging NFRCs, the fashion industry is transitioning toward sustainable practices without compromising on style or quality, ultimately paving the way for a future where creativity and environmental responsibility coexist harmoniously. This shift not only benefits designers and consumers but also contributes to reducing the ecological footprint of the fashion industry, aligning it with global efforts toward sustainability.

3 Sustainability Metrics of Natural Fiber-Reinforced Composites

3.1 Environmental Benefits

The adoption of natural fiber-reinforced composites (NFRCs) in the fashion industry represents a significant step toward achieving broader sustainability goals, as it directly addresses some of the most pressing environmental concerns of our time. One of the primary benefits of NFRCs is their potential to significantly reduce carbon emissions during production [22]. Synthetic fibers, such as polyester and nylon, are derived from petroleum-based sources, which are not only nonrenewable but also energy-intensive to produce. The extraction and processing of these materials contribute heavily to greenhouse gas emissions, exacerbating climate change. In contrast, natural fibers like bamboo, flax, hemp, and jute are biodegradable and have a much lower environmental impact. Studies indicate that replacing synthetic fibers with natural ones can reduce greenhouse gas emissions by up to 50% during the production phase. This reduction is primarily due to the fact that natural fibers grow through photosynthesis, a process that absorbs carbon dioxide from the atmosphere, thereby offsetting some of the emissions generated during harvesting and processing [23].

In addition to reducing carbon footprints, NFRCs offer notable advantages in terms of energy efficiency. The manufacturing processes for natural fibers typically consume far less energy compared to their synthetic counterparts. For instance, producing a kilogram of polyester requires substantially more energy than producing the same amount of hemp or flax fiber. This disparity arises because natural fibers often undergo simpler mechanical treatments rather than complex chemical reactions, which are common in the synthesis of synthetic materials. By shifting toward NFRCs, the fashion industry can decrease its reliance on fossil fuels and transition to more sustainable practices, ultimately conserving valuable resources and mitigating environmental degradation [24].

Another critical environmental benefit of NFRCs is their enhanced recyclability and potential for repurposing. Unlike many synthetic textiles, which are difficult to recycle and often end up in landfills, natural fiber composites can be broken down and reused at the end of their lifecycle. For example, bamboo or hemp-based composites can be shredded and reprocessed into new materials, minimizing waste generation and promoting a circular economy. Furthermore, even when NFRCs are discarded, they decompose naturally without releasing harmful microplastics into the environmental growing concern associated with synthetic fabrics. By embracing NFRCs, the fashion industry can address issues related to pollution, resource depletion, and waste management while aligning itself with global efforts to combat climate change. This shift not only supports ecological balance but also enhances the industry's reputation as a responsible steward of the planet, meeting the rising demand for sustainable products among environmentally conscious consumers [25] (Table 2).

Table 2 Compares the energy required to produce various types of fibers (measured in megajoules per kilogram, MJ/kg)

Fiber type	Energy consumption (MJ/kg)	Notes
Synthetic fibers		
Polyester	~125	Petroleum extraction + high-heat processing
Nylon	~150	Energy-intensive polymerization
Acrylic	~170	Fossil fuels + chemical processes
Natural fibers		
Cotton (conventional)	~55	Irrigation, fertilizers, pesticides
Cotton (organic)	~30	Lower energy (no synthetic chemicals)
Wool	~60	Sheep farming, feed, and processing
Hemp	~15	Low-input crop, minimal processing
Silk	~180	Labor-intensive cultivation and processing

3.2 *Life Cycle Assessment (LCA)*

Life Cycle Assessment (LCA) is a systematic methodology used to evaluate the environmental impacts of a product, material, or system throughout its entire life cycle. This process involves analyzing every stage, from raw material extraction and processing to manufacturing, use, and eventual disposal or recycling. LCA provides a holistic view of sustainability by considering energy consumption, greenhouse gas emissions, water usage, chemical inputs, waste generation, and other environmental factors [26].

A comprehensive LCA reveals that natural fiber-reinforced composites (NFRCs), such as flax-based composites, outperform traditional materials like cotton, polyester, or synthetic alternatives across multiple stages. For example, during the raw material extraction phase, natural fibers like flax require significantly fewer resources compared to synthetic counterparts derived from petroleum. Flax cultivation typically demands less water and relies on fewer chemical fertilizers and pesticides than crops like cotton, which are notorious for their high water consumption and heavy pesticide use. Similarly, flax production emits fewer greenhouse gases compared to the energy-intensive processes required to manufacture polyester [27].

In the processing stage, NFRCs also demonstrate advantages. The transformation of flax into composite materials often involves lower energy inputs and reduced reliance on harmful chemicals compared to refining petroleum-based synthetics. Additionally, flax-based composites exhibit excellent durability and mechanical properties, making them suitable for various applications while reducing the need for frequent replacements, a key factor in minimizing long-term environmental impact during the "use" phase. At the end-of-life stage, NFRCs offer better recyclability and biodegradability compared to nonrenewable materials like polyester. When disposed of, flax-based composites decompose naturally without leaving behind persistent microplastics or toxic residues, unlike many synthetic alternatives. Overall, LCAs highlight how NFRCs contribute to more sustainable practices by lowering resource consumption, reducing pollution, and mitigating ecological footprints across all life cycle stages [28].

4 Real-World Applications

4.1 *Apparel*

Natural fiber-reinforced composites (NFRCs) are revolutionizing the fashion industry by offering innovative, sustainable, and versatile materials that push the boundaries of design in apparel. These composites, made from natural fibers such as bamboo, flax, hemp, or jute embedded in a polymer matrix, combine strength, flexibility, and lightweight properties, making them ideal for avant-garde fashion creations. The use of NFRCs in apparel is not only transforming aesthetics but also addressing growing concerns about sustainability within the fashion industry [29].

One of the most exciting aspects of NFRCs in apparel is their ability to enable bold, sculptural silhouettes and intricate architectural details. Traditional textiles often lack the structural integrity required to create three-dimensional forms or maintain complex shapes. However, NFRCs provide the necessary rigidity while remaining lightweight, allowing designers to experiment with unconventional designs that were previously impossible. For example, a recent high-profile runway collection showcased dresses constructed from bamboo-reinforced panels. These garments featured dramatic geometric patterns and flowing yet structured forms, demonstrating how NFRCs can elevate fashion into an art form. Bamboo, known for its rapid renewability and minimal environmental impact, serves as both a functional and eco-friendly material choice [30].

Beyond their aesthetic appeal, NFRCs offer practical advantages for apparel design. Their durability ensures that garments retain their shape over time, even under stress or movement, which is particularly beneficial for performance wear or statement pieces intended for long-term use. Additionally, the lightweight nature of these composites enhances comfort, making them suitable for everyday clothing as well as high-fashion applications. The integration of NFRCs into apparel also aligns with the growing demand for sustainable fashion. As consumers become more conscious of the environmental footprint of their clothing choices, designers are seeking alternatives to synthetic fabrics derived from fossil fuels. Natural fiber composites present a compelling solution, as they are biodegradable, renewable, and often require fewer chemical inputs during production compared to conventional textiles like polyester or nylon. This shift toward sustainable materials helps reduce waste and pollution throughout the lifecycle of the garment [31].

Moreover, the versatility of NFRCs allows for customization and innovation in texture, finish, and functionality. Designers can manipulate these materials to achieve unique surface effects, such as glossy finishes or textured weaves, further expanding creative possibilities. In summary, NFRCs are proving to be a game-changer in the world of apparel, enabling groundbreaking designs while promoting sustainability and reducing environmental harm. By merging cutting-edge technology with eco-conscious practices, NFRCs are paving the way for a new era of fashion that is both visually stunning and ethically responsible.

4.2 Accessories

Natural fiber-reinforced composites (NFRCs) are making significant inroads into the world of fashion accessories, offering a compelling blend of durability, aesthetic appeal, and sustainability. Items such as handbags, belts, and jewelry are benefiting immensely from the unique properties of NFRCs, which enable designers to create innovative, stylish, and environmentally conscious products. These composites, derived from natural fibers like hemp, flax, bamboo, or jute embedded in polymer matrices, provide a robust yet lightweight material that can be molded into intricate shapes while maintaining structural integrity. This versatility makes them ideal for crafting high-quality accessories that meet both functional and aesthetic demands [32].

One of the standout advantages of NFRCs in accessories is their ability to combine sleek design with eco-consciousness. For instance, a luxury brand recently launched a collection of handbags made from hemp-reinforced composites, showcasing how these materials can elevate everyday items into premium, sustainable fashion statements. Hemp, known for its strength, renewability, and low environmental impact, serves as an excellent base material for creating durable handbags that retain their shape over time. The composite's smooth finish and customizable textures allow designers to achieve a polished, modern look that appeals to discerning consumers seeking both style and sustainability [33].

Beyond handbags, NFRCs are also transforming other accessories such as belts and jewelry. Belts made from flax or bamboo composites offer a unique combination of flexibility and rigidity, ensuring they remain sturdy while providing comfort during wear. Similarly, jewelry crafted from NFRCs allows designers to experiment with bold shapes and intricate details, creating pieces that are lightweight yet striking. These materials can be dyed, textured, or finished in various ways to mimic traditional materials like leather, metal, or plastic, while offering a more sustainable alternative. The use of NFRCs in accessories aligns perfectly with the growing consumer demand for eco-friendly products. Unlike conventional materials such as leather, PVC, or synthetic resins, which often involve resource-intensive production processes and contribute to pollution, NFRCs are biodegradable, renewable, and require fewer chemical inputs. This shift not only reduces the environmental footprint of accessories but also positions brands as leaders in sustainable innovation.

4.3 Footwear

Natural fiber-reinforced composites (NFRCs) are increasingly being adopted in the footwear industry, where they enhance performance, durability, and sustainability without compromising on style. These innovative materials, derived from natural fibers like flax, bamboo, hemp, or jute, offer unique properties that make them ideal for various components of shoes, including midsoles, outsoles, and structural reinforcements. By integrating NFRCs into footwear design, brands are able to create products that cater to both functional needs and aesthetic preferences, while also addressing the growing demand for environmentally responsible solutions [34].

One of the key applications of NFRCs in footwear is in the development of midsoles, which are critical for providing cushioning, support, and comfort. Flax-based composites, for instance, have emerged as a popular choice for midsole construction due to their lightweight nature and excellent shock-absorbing properties. These materials effectively distribute pressure across the foot, reducing strain during walking or running. Additionally, flax fibers are renewable and biodegradable, making them a more sustainable alternative to traditional petroleum-based foams commonly used in midsoles. This shift not only improves the environmental footprint of footwear but also enhances the overall user experience by delivering superior performance [35].

In addition to midsoles, bamboo-reinforced composites are being utilized to add structural integrity to footwear. Bamboo fibers, known for their strength and flexibility, are incorporated into components such as heel counters or arch supports, ensuring the shoe maintains its shape over time. This reinforcement is particularly beneficial for athletic footwear, where durability and stability are paramount. Moreover, bamboo‘s rapid renewability and minimal resource requirements further underscore its appeal as an eco-friendly material. Several forward-thinking brands, such as Allbirds, have successfully integrated NFRCs into their product lines, launching sneakers and sandals that combine cutting-edge technology with stylish designs. These products demonstrate how natural fiber composites can be molded into sleek, modern silhouettes while maintaining functionality. For example, flax-based midsoles paired with bamboo reinforcements result in shoes that are not only visually appealing but also highly durable and comfortable [36].

The adoption of NFRCs in footwear aligns with the industry’s push toward sustainability, as these materials reduce reliance on synthetic polymers and nonrenewable resources. By incorporating NFRCs, brands are able to meet consumer expectations for eco-conscious products without sacrificing performance or aesthetics. In summary, NFRCs are revolutionizing footwear by offering a harmonious blend of style, functionality, and environmental responsibility, paving the way for a more sustainable future in fashion.

5 Challenges and Emerging Solutions

5.1 Scalability

While natural fiber-reinforced composites (NFRCs) have demonstrated immense potential across industries such as apparel, accessories, and footwear, scaling their production to meet mass-market demands remains a significant challenge. The primary obstacles stem from the labor-intensive processes involved in sourcing, processing, and manufacturing these materials, which can hinder their widespread adoption. However, emerging technological advancements, particularly in automation and additive manufacturing, are beginning to address these issues, paving the way for more efficient and scalable production of NFRCs [37].

One of the key challenges in scaling NFRC production lies in the extraction and preparation of natural fibers. Unlike synthetic materials, which are often derived from standardized industrial processes, natural fibers like flax, hemp, or bamboo require careful harvesting, cleaning, and treatment to ensure consistent quality. These steps are often manual or semi-automated, making them time-consuming and costly. Additionally, variations in fiber properties due to differences in growing conditions or processing techniques can lead to inconsistencies in the final composite material, posing further challenges for large-scale manufacturing. Another hurdle is the integration of natural fibers into polymer matrices to create composites. This process typically involves mixing, molding, and curing, which can be energy-intensive and require specialized equipment. Current methods often rely on traditional manufacturing techniques that are not optimized for high-volume production, limiting the ability to produce NFRCs at scale [38].

Despite these challenges, advancements in automation and additive manufacturing are offering promising solutions. Automation technologies, such as robotic systems and machine learning algorithms, are being developed to streamline the harvesting, processing, and fabrication of natural fibers. These innovations reduce reliance on manual labor, improve efficiency, and enhance consistency in material quality. For example, automated fiber extraction machines can now process raw materials faster and with greater precision, ensuring uniformity in fiber length and texture. Additive manufacturing, or 3D printing, is another game-changing technology that is revolutionizing NFRC production. By enabling layer-by-layer construction of complex shapes and structures, 3D printing reduces material waste and allows for the customization of NFRC components. This approach is particularly beneficial for creating intricate designs in fashion, footwear, and accessories, where traditional manufacturing methods may fall short. Moreover, 3D printing can accommodate a wide range of natural fibers, making it easier to experiment with different combinations and optimize material performance [39].

5.2 Consumer Acceptance

While natural fiber-reinforced composites (NFRCs) offer numerous environmental and functional advantages, achieving widespread consumer acceptance remains a critical hurdle. Many consumers are still unfamiliar with NFRCs or may harbor misconceptions about their performance, durability, or aesthetic appeal compared to traditional materials like synthetic fibers or leather. To overcome these barriers, educating consumers about the benefits of NFRCs is essential. Strategic marketing campaigns that emphasize sustainability, cultural significance, and the innovative nature of these materials can play a pivotal role in shifting perceptions and driving adoption [40].

One of the primary challenges in gaining consumer acceptance is the perception that sustainable materials might compromise on quality or style. For instance, some consumers may assume that products made from natural fibers lack the durability or sophistication of conventional materials. To counter this, brands must highlight the unique properties of NFRCs, such as their strength, lightweight nature, and versatility in design. Demonstrating how NFRCs can be used to create high-performance, stylish, and long-lasting products, whether in apparel, accessories, or footwear—can help dispel doubts and build trust [41].

Marketing campaigns that focus on the sustainability of NFRCs are particularly effective in appealing to environmentally conscious consumers. By emphasizing the renewable nature of natural fibers, their lower carbon footprint, and their biodegradability, brands can position NFRC-based products as a responsible choice for those seeking to reduce their environmental impact. For example, showcasing the lifecycle benefits of flax-based composites or bamboo reinforcements in comparison to synthetic alternatives can resonate with eco-aware audiences and encourage them to make more sustainable purchasing decisions. In addition to sustainability, highlighting the cultural significance of natural fibers can further enhance consumer acceptance. Many natural fibers, such as hemp, flax, and bamboo, have deep-rooted histories in various cultures and are often associated with craftsmanship and heritage. Brands can leverage these narratives to create emotional connections with consumers, positioning NFRCs not just as innovative materials but also as a celebration of tradition and cultural identity. Collaborations with local artisans or storytelling through marketing materials can amplify this message. Finally, transparency and education are key to fostering trust. Brands should provide clear information about the sourcing, production processes, and benefits of NFRCs through labeling, digital content, and in-store experiences. Interactive campaigns, such as workshops or virtual tours of production facilities, can engage consumers and deepen their understanding of these materials [42].

5.3 Cost-Effectiveness

While natural fiber-reinforced composites (NFRCs) present numerous environmental and functional advantages, their cost-effectiveness remains a critical consideration for both manufacturers and consumers. Initially, the production costs of NFRCs can be higher compared to traditional materials like synthetic fibers or petrochemical-based composites. This is largely due to the labor-intensive processes involved in harvesting, processing, and manufacturing natural fibers, as well as the current limited scale of production. However, as technological advancements continue to evolve and economies of scale are achieved, the costs associated with NFRCs are expected to decrease significantly over time, making them more competitive in the market. One of the primary factors contributing to the higher initial costs of NFRCs is the complexity of integrating natural fibers with polymer matrices. The process often requires specialized equipment and precise control over material properties to ensure consistency and performance. Additionally, the supply chain for natural fibers is not yet as streamlined as that of synthetic materials, which have been optimized over decades of industrial use. These inefficiencies can lead to higher production expenses, particularly in the early stages of adoption.

Despite these challenges, several factors are expected to drive down costs in the long term. First, economies of scale will play a crucial role as demand for NFRCs grows and production volumes increase. Larger-scale operations will reduce per-unit costs by spreading fixed expenses across a greater number of products. For instance, as more brands incorporate NFRCs into their offerings—whether in apparel, accessories, footwear, or other industries—the increased demand will incentivize investments in more efficient manufacturing processes and infrastructure. Technological improvements are also expected to enhance cost-effectiveness. Innovations in automation, additive manufacturing, and bio-based resin development are streamlining production and reducing reliance on labor-intensive methods. For example, bio-based resins, which are derived from renewable resources such as plant oils or starches, offer significant cost-saving opportunities by reducing dependency on volatile and increasingly expensive petrochemicals. These resins not only lower material costs but also align with sustainability goals, further enhancing the value proposition of NFRCs. Moreover, the long-term economic benefits of NFRCs should not be overlooked. Their durability, lightweight nature, and reduced environmental impact can lead to savings in areas such as transportation, maintenance, and waste management. For instance, lighter materials reduce fuel consumption during shipping, while biodegradable composites minimize disposal costs.

6 Conclusion

The integration of natural fiber-reinforced composites (NFRCs) into the fashion industry represents a transformative step toward sustainability, innovation, and cultural preservation. By merging advanced material science with traditional craftsmanship, NFRCs offer a unique opportunity to address some of the most pressing challenges facing the fashion industry today, including environmental degradation, resource depletion, and the loss of cultural heritage. This chapter has explored the multifaceted potential of NFRCs, highlighting their environmental benefits, functional versatility, and capacity to align with consumer demands for eco-conscious and culturally significant products. As demonstrated through real-world applications in apparel, accessories, and footwear, NFRCs are not only capable of meeting contemporary design and performance standards but also of pushing creative boundaries, enabling designers to craft avant-garde pieces that blend aesthetics with sustainability.

Life cycle assessments (LCAs) have underscored the environmental advantages of NFRCs, revealing their lower carbon footprints, reduced energy consumption, and enhanced recyclability compared to synthetic alternatives. These materials, derived from renewable resources like flax, hemp, and bamboo, exemplify a harmonious balance between ecological responsibility and industrial innovation. However, challenges such as scalability, cost-effectiveness, and consumer acceptance remain significant barriers to widespread adoption. Emerging solutions, including advancements in automation, additive manufacturing, and bio-based resins, offer promising pathways to overcoming these hurdles, paving the way for more efficient and economically viable production processes.

Ultimately, NFRCs embody the intersection of tradition and innovation, proving that style and sustainability can coexist seamlessly in modern fashion. By embracing these materials, the industry can transition toward practices that not only mitigate environmental harm but also celebrate diverse cultural identities. As consumer awareness grows and technological advancements continue to evolve, NFRCs hold the potential to redefine the future of fashion, fostering a world where creativity, cultural heritage, and ecological stewardship converge. This vision underscores the urgent need for collaboration among designers, manufacturers, and policymakers to accelerate the adoption of NFRCs and ensure a sustainable legacy for generations to come.

References

1. Karim, M. A., Abdullah, M. Z., Deifalla, A. F., Azab, M., & Waqar, A. (2023). An assessment of the processing parameters and application of fibre-reinforced polymers (FRPs) in the petroleum and natural gas industries: A review. *Results in Engineering, 18*(April), 101091. https://doi.org/10.1016/j.rineng.2023.101091
2. Karim, M. A., Abdullah, M. Z., & Ahmed, T. (2021). An overview: The processing methods of fiber-reinforced polymers (FRPS). *International Journal of Mechanical Engineering and Technology, 12*(2), 10–24. https://doi.org/10.34218/IJMET.12.2.2021.002
3. Karim, M. A., Abdullah, M. Z., Waqar, A., Deifalla, A. F., Ragab, A. E., & Khan, M. (2023). Analysis of the mechanical properties of the single layered braid reinforced thermoplastic pipe (BRTP) for oil & gas industries. *Results in Engineering, 20*, 101483. https://doi.org/10.1016/j.rineng.2023.101483
4. Karim, M. A. (2024). *Hoop, stiffness and axial tensile properties of the single layered braid-reinforced thermoplastic pipe for oil and gas applications*. Universiti Teknologi PETRONAS. https://doi.org/10.13140/RG.2.2.26385.03685
5. Rajak, D., Pagar, D., Menezes, P., & Linul, E. (2019). Fiber-reinforced polymer composites: Manufacturing, properties, and applications. *Polymers (Basel), 11*(10), 1667. https://doi.org/10.3390/polym11101667
6. Jayan, J. S., Appukuttan, S., Wilson, R., Joseph, K., George, G., & Oksman, K. (2021). An introduction to fiber reinforced composite materials. In *Fiber reinforced composites* (pp. 1–24). Elsevier.
7. Vallittu, P. K. (2018). An overview of development and status of fiber-reinforced composites as dental and medical biomaterials. *Acta Biomaterialia Odontologica Scandinavica, 4*(1), 44–55.
8. Lotfi, A., Li, H., Dao, D. V., & Prusty, G. (2021). Natural fiber–reinforced composites: A review on material, manufacturing, and machinability. *Journal of Thermoplastic Composite Materials, 34*(2), 238–284.
9. Vallittu, P. K. (2015). High-aspect ratio fillers: Fiber-reinforced composites and their anisotropic properties. *Dental Materials, 31*(1), 1–7.
10. Çevik, M., & Diambu, A. N. (2024). *Advancing sustainable development goals through the use of biocomposites for a greener future*. Scientific Research Reports.
11. Skosana, S. J., Khoathane, C., & Malwela, T. (2024). Driving towards sustainability: A review of natural fiber reinforced polymer composites for eco-friendly automotive light-weighting. *Journal of Thermoplastic Composite Materials, 38*(2), 754–780.
12. Abdullah, C. C., & Karim, M. A. (2024). Optimization of solar power plant with variation of solar reflector angles and use of passive cooling integrated internet of things. *Journal of Advanced Research in Fluid Mechanics Thermal Science, 124*(1), 233–248. https://doi.org/10.37934/arfmts.124.1.233248

13. Cholish, C., et al. (2024). Performance measurement of peltier element design using solar test simulator concerning light and temperature parameters. *Journal of Advanced Research in Fluid Mechanics Thermal Sciences, 123*(2), 231–243.
14. Thyavihalli Girijappa, Y. G., Mavinkere Rangappa, S., Parameswaranpillai, J., & Siengchin, S. (2019). Natural fibers as sustainable and renewable resource for development of eco-friendly composites: A comprehensive review. *Frontiers in Materials, 6*, 226.
15. Harahap, P., et al. (2024). Power conversion from solar panels using a 3000-watt inverter. *Journal of Advanced Research in Fluid Mechanics and Thermal Sciences, 124*(2), 260–272.
16. Bismarck, A., Mishra, S., & Lampke, T. (2005). Plant fibers as reinforcement for green composites. In *Natural fibers, biopolymers, and biocomposites* (pp. 52–128). CRC Press.
17. de Beus, N., Carus, M., & Barth, M. (2019). *Carbon footprint and sustainability of different natural fibres for biocomposites and insulation material: Study providing data for the automotive and insulation industry*. Nova-Institut GmbH.
18. Islam, T., et al. (2024). Advancements and challenges in natural fiber-reinforced hybrid composites: A comprehensive review. *SPE Polymers, 5*(4), 481–506.
19. Mohammed, M., et al. (2024). A review on the advancement of renewable natural fiber hybrid composites: Prospects, challenges, and industrial applications. *Journal Renewable Materials, 12*(7), 1237–1290.
20. Lubis, S., Napitupulu, F. H., Abdullah, I., Sitorus, T. B., & Karim, M. A. (2024). Study of comparison heat energy disposal in hollow roof cooling water tank using natural material. *Journal of Advanced Research in Applied Sciences and Engineering Technology, 55*(2), 14–42.
21. Ram Kishore, S., et al. (2024). Natural fiber biocomposites via 4D printing technologies: A review of possibilities for agricultural bio-mulching and related sustainable applications. *Progress in Additive Manufacturing, 9*(1), 37–67.
22. Abdollahiparsa, H., Shahmirzaloo, A., Teuffel, P., & Blok, R. (2023). A review of recent developments in structural applications of natural fiber-reinforced composites (NFRCs). *Composites and Advanced Materials, 32*, 26349833221147540.
23. Thapliyal, D., et al. (2023). Natural fibers composites: Origin, importance, consumption pattern, and challenges. *Journal of Composites Science, 7*(12), 506.
24. Singh, A., & Yadav, B. P. (2024). Sustainable innovations and future prospects in construction material: A review on natural fiber-reinforced cement composites. *Environmental Science and Pollution Research, 31*, 62549–62587.
25. Tyagi, A., Meena, A., Bhandari, M., Shahid, M., Gupta, B. L., & Meena, S. L. (2024). A critical review on additive manufacturing of sustainable NFRCs: Processing and applications. *Green Materials*. https://doi.org/10.1680/jgrma.24.00126
26. Nwodo, M. N., & Anumba, C. J. (2019). A review of life cycle assessment of buildings using a systematic approach. *Building and Environment, 162*, 106290.
27. Islam, S., et al. (2025). Thermoset and thermoplastic polymer composites reinforced with flax fiber: Properties and application—A review. *SPE Polymers, 6*(1), e10172.
28. Prem Kumar, R., Muthukrishnan, M., & Felix Sahayaraj, A. (2023). Effect of hybridization on natural fiber reinforced polymer composite materials–a review. *Polymer Composites, 44*(8), 4459–4479.
29. Tyagi, A., Gupta, B. L., Mukhija, L., Bhandari, M., Meena, A., & Meena, S. L. (2024). Mechanical and tribological behavior of NFRCs: A comprehensive analysis of properties and applications. *Green Materials*. https://doi.org/10.1680/jgrma.24.00113
30. Tasgin, Y., Demircan, G., Kandemir, S., & Acikgoz, A. (2024). Mechanical, wear and thermal properties of natural fiber-reinforced epoxy composite: Cotton, sisal, coir and wool fibers. *Journal of Materials Science, 59*(24), 10844–10857.
31. Repon, M. R., Islam, T., Islam, T., & Alim, M. A. (2024). Manufacture of polymer composites from plant fibers. In *Plant biomass derived materials sources, extractions, and applications* (pp. 363–388). Wiley-VCH.
32. Saleh Alghamdi, S., John, S., Roy Choudhury, N., & Dutta, N. K. (2021). Additive manufacturing of polymer materials: Progress, promise and challenges. *Polymers (Basel), 13*(5), 753.

33. Prabhu, R., Mendonca, S., Bellairu, P. K., DSouza, R. C., & Bhat, T. (2024). Investigating the impact of TiO_2 filler on abrasive wear characteristics of bamboo fiber-reinforced epoxy composites using the Taguchi method. *World Journal of Engineering, 22*(2), 237–249.
34. Castro-Franco, A. D., Siqueiros-Hernández, M., García-Angel, V., Mendoza-Muñoz, I., Vargas-Osuna, L. E., & Magaña-Almaguer, H. D. (2024). A review of natural fiber-reinforced composites for lower-limb prosthetic designs. *Polymers (Basel), 16*(9), 1293.
35. Sathish, M., Radhika, N., Venuvanka, N., & Rajeshkumar, L. (2024). A review on sustainable properties of plant fiber-reinforced polymer composites: Characteristics and properties. *Polymer International, 73*(11), 887–943.
36. Possidónio, C., Farias, A. R., Domingos, S., Cruz, B., Luís, S., & Loureiro, A. (2025). Exploring supply-side barriers for commercialization of new biopolymer production technologies: A systematic review. *Sustainability, 17*(3), 820.
37. Simões, S. (2024). High-performance advanced composites in multifunctional material design: State of the art, challenges, and future directions. *Materials (Basel), 17*(23), 5997.
38. Sathish, S., et al. (2021). A review of natural fiber composites: Extraction methods, chemical treatments and applications. *Materials Today Proceedings, 45*, 8017–8023.
39. Navas, E., Fernández, R., Sepúlveda, D., Armada, M., & Gonzalez-de-Santos, P. (2021). Soft grippers for automatic crop harvesting: A review. *Sensors, 21*(8), 2689.
40. Harris, F., Roby, H., & Dibb, S. (2016). Sustainable clothing: Challenges, barriers and interventions for encouraging more sustainable consumer behaviour. *International Journal of Consumer Studies, 40*(3), 309–318.
41. Mylsamy, B., Shanmugam, S. K. M., Aruchamy, K., Palanisamy, S., Nagarajan, R., & Ayrilmis, N. (2024). A review on natural fiber composites: Polymer matrices, fiber surface treatments, fabrication methods, properties, and applications. *Polymer Engineering and Science, 64*(6), 2345–2373.
42. Banea, M. D., De Queiroz, H. F. M., Neto, J. S. E. S., & Cavalcanti, D. K. K. (2024). Recent advances in development, characterization and joining of new sustainable materials. *Materials Research Proceedings, 41*, 530–539.

Heritage Fashion: A Catalyst for Sustainability

R. Surjit, R. Akshya, R. Akshatha, and P. Varshitha

1 Introduction to Heritage and Sustainability in Fashion

In South Asian culture, the textiles are used widely, other than in the apparel industry, such as religious clothing, decorative items, festival clothing, festival decorations, and so many. As the saying "old is gold" suggests making of these by heritage brands expresses the respect shown to the good through the process, and ethics are followed for the well-being of the artisan and the environment. Heritage and sustainable fashion are practiced by many societies around the world. For example, Indonesian tribes produce handmade textiles using natural dyes and traditional techniques [1]. They weave fabrics with palm leaf slivers for shine and intricate patterns. The T'boli, known as "Dream Weavers," use dark vegetable dyes on wild banana fiber, employing the ikat method for unique contrasts.

1.1 Definition of Heritage Fashion

Heritage fashion refers to clothing and accessories that draw inspiration from historical styles, traditional craftsmanship, and cultural legacies, blending them with contemporary aesthetics to create timeless pieces. This approach highlights real materials, timeless designs, and a respect for cultural narratives as an alternative to the rapid cycle of fast fashion. A supreme example of heritage fashion is the Hanfu

R. Surjit (✉) · R. Akshya · R. Akshatha · P. Varshitha
Department of Fashion Technology, PSG College of Technology,
Coimbatore, Tamil Nadu, India
e-mail: rst.fashion@psgtech.ac.in; 22h203@psgtech.ac.in; 22h202@psgtech.ac.in;
22h245@psgtech.ac.in

S. S. Muthu (ed.), *Cultural Sustainability in Fashion*, Sustainable Textiles:
Production, Processing, Manufacturing & Chemistry,
https://doi.org/10.1007/978-3-032-03207-2_4

Movement in China, which aims to restore and revive traditional Han Chinese clothing in a modern way. Hanfu enthusiasts see these items as an opportunity to connect with their cultural heritage and symbolize their commitment to traditional values by asserting the cultural and historical significance of these garments. The movement also implicitly acknowledges the role of dress in expressing cultural identity and preserving cultural narratives. Another example of heritage fashion includes the use of ethnic designs in fashion today. An example is the Li brocade, which is one of the China's oldest textile heritages. The use of this kind of ethnic design is important in keeping traditional crafts alive and bringing them to wider audiences in the modern era of fashion. Heritage fashion offers garments with stories and connections to specific cultures or historical periods, reflecting a growing appreciation for craftsmanship and the narratives embedded in clothing. Heritage fashion often emphasizes quality and durability characteristics that contrast sharply with the disposable nature of much contemporary clothing. Brands associated with heritage fashion typically produce items designed to last, appealing to consumers who value sustainability and wish to invest in pieces with longevity. This focus on enduring quality aligns with the principles of slow fashion, promoting thoughtful consumption and a deeper connection to one's wardrobe. Heritage fashion always serves as a bridge between the past and present, allowing the generations to carry on the historical and cultural significance. The consumers, along with celebrating traditional craftsmanship, also contribute to the preservation of cultural narratives by embracing the heritage fashion.

1.2 Heritage Fashion Brands

These fashion brands have a relook at the history and create timeless collections reflecting the heritage of the particular place and the world. There are many heritage brands worldwide. In this section, some of the heritage brands are highlighted.

1.2.1 House of Wandering Silk

House of Wandering Silk, founded in 2011 by Australian humanitarian Katherine Neumann in Delhi, began as a one-woman brand inspired by Gandhi's belief that "there is no beauty in the finest cloth if it brings hunger and unhappiness." The brand partners with artisans across India to preserve traditional textile crafts through handmade, zero-waste, small-batch clothing, and accessories.

Committed to ethical production, the brand empowers marginalized women through fair wages and sustainable livelihoods. Prioritizing transparency, it brings the people and stories behind every piece to the forefront. Although a carefully curated selection is available online, the brand's primary focus is wholesale, with prominent partners such as the Victoria & Albert Museum Store in London.

1.2.2 My Sleeping Gypsy

My Sleeping Gypsy, established by Katya Hermann in the Ukrainian Revolution, is now in the hands of five women who have different backgrounds but a similar vision of mindful consumerism by way of sustainable design. My Sleeping Gypsy is all about sustainability, culture, and heritage, encouraging slow fashion through eternity pieces created to last.

Made from local hemp and linen, every item has detailed, handmade embroidery produced through collaborative work with international artists. These exclusive ornaments are used as a symbolic language and convey the theme of harmony and balance. Available in more than 60 retail outlets in 15 nations, My Sleeping Gypsy wants to empower the wearer with significant, responsibly produced apparel that promotes conscientious decisions and honors craftsmanship.

1.2.3 Buki Akomolafe

Buki Akomolafe is a Berlin fashion brand that is based on the designer‘s German-Nigerian heritage. Drawing inspiration from her childhood on the continent of Africa and an admiration for frugal living, the label combines German tailoring with Nigerian-West African workmanship to produce elegant, androgynous women's clothing.

Emphasizing sustainability, Buki Akomolafe employs ethically obtained materials, including organic cotton, hemp-silk, and African wax prints purchased from the Netherlands. All pieces are handcrafted in Berlin to minimize environmental footprint and express devotion to quality, tradition, and cross-cultural exchange.

1.2.4 Filanda N.18

Filanda N.18 is a fashion design studio established by Francesca to revive and reinterpret the world's textile heritage in the form of limited-edition ready-to-wear collections. Every collection investigates traditional methods—from Uzbek ikat and Venetian lace to Bengali kantha and Japanese boro—through extensive research, artisanal collaboration, and cultural narrative.

The design process includes researching the origins of every craft, traveling to artisans all over the world, and creating pieces in collaboration with Italian workshops, Indian women-led communities, and Venice lace restorers. From deadstock luxury Northern Italian fabrics, Filanda N.18 produces garments that pay homage to craftsmanship, culture, and sustainability, safeguarding lost textile traditions and their stories.

1.2.5 Effei

Effei is a progressive, activist fashion brand dedicated to sustainability and cultural heritage. Each piece is lovingly handmade using fabrics sourced from five continents, with traditional techniques that honor the best of global heritage and local economies.

The line features handcrafted waistcoats and tote bags from places, such as Bali, Australia, Burkina Faso, Mexico, and Venice, in addition to the company's best-selling organic cotton T-shirts—the Skyline Collection, hand-personalized with city skylines, and the Wanderlust Collection, inspired by travel and cultural curiosity.

Inspired by the motto "Wearing cultures," Effei places ecological and social responsibility at the forefront of each step—from certified organic supplies and reduced-impact dyes to plantable, recycled tags and eco-friendly packaging. The company maintains transparency through direct artisan partnership, providing customers with a conscious and intentional fashion experience.

All the above brands and many more heritage fashion brands focus on sustainability and thereby they create products that are sustainable for a longer period of time. This shows that sustainability is being embraced by everyone and it is the need of the hour.

1.3 Importance of Sustainability in Fashion

The fashion industry exists as one of the most polluting industries in the world. It is due to the usage of tons of water, excessive usage of energy and other non-renewable resources, and the production of excess carbon dioxide, harmful chemicals, and waste. The fashion industry involves the process of sourcing raw materials, processing, and production, and in all these, sustainability acts as an important aspect. Therefore, the fashion industry must be responsible to the community and environment by ensuring less negative impact. The fashion industry has many negative impacts on the environment, considering the process of consumer disposal. An estimation of over 20 million tons of textiles is dumped annually in China and 2000 tons of textiles in Malaysia. Fashion waste accounts for roughly 7% of solid fashion waste, 20% of global wastewater in the world, and is responsible for 10% of the world's carbon emissions. The amount of clothing produced is increasing yearly; at the same time, unused clothing is rising. These unused clothes are sent to disposal sites, which contributes to an increase in the release of methane and other hazardous gases during their decomposition. This also affects the soil purity, since most textiles are made of synthetic materials. Additionally, consumers do not seem to be aware that sustainable fashion does not necessarily involve higher prices. Considering the overall effects on the fashion industry, it is imperative to understand the importance of sustainability in fashion by reducing environmental pollution, increasing consumer convenience, reducing costs, and increasing societal development. Sustainable practices in the fashion industry have many advantages. When the clothes are produced sustainably, it can reduce environmental pollution, fulfill consumers' wants at a reduced cost. It can reduce the release of carbon dioxide, methane, and other hazardous gases. The rise of negative impacts results in an increase in awareness of sustainable practices and the importance of the environment [2].

2 Traditional Craftsmanship in Heritage Fashion

Traditional craftsmanship plays a vital role in preserving and embracing cultural identity while promoting sustainability. One of the most significant traditional textile techniques is handweaving, which embodies the blend of artistry, eco-conscious production, and community-driven heritage. Various indigenous communities across the world have been practicing and sustaining this craft through generations, creating textiles that reflect cultural narratives, environmental ethics, and social resilience.

2.1 Handweaving Techniques

2.1.1 Handweaving and Sustainability

Handweaving techniques align with sustainable fashion by minimizing resource exploitation and reducing environmental footprints. Unlike mechanized textile production, which relies heavily on synthetic fibers and energy-intensive processes, handwoven textiles emphasize:

- **Use of Natural Fibers:** Materials such as organic cotton, wool, and silk (including Eri silk) are used, which are biodegradable and environmentally friendly.
- **Eco-friendly Dyeing Methods:** Many handloom traditions employ plant-based dyes derived from turmeric, madder root, and indigo, reducing chemical waste and water pollution.
- **Low Energy Consumption:** Since handlooms require no electricity, they offer an energy-efficient alternative to power looms.

One prominent example of sustainable handweaving is Eri silk weaving, known as "peace silk," because it allows silkworms to complete their life cycle before the fibers are extracted. This ethical approach, combined with the silk's thermal properties, makes it a preferred choice for sustainable textiles [3].

2.1.2 Cultural and Artistic Significance

Handwoven textiles are deeply intertwined with cultural heritage, often carrying symbolic motifs and regional identities. Indigenous craftsmanship reflects a community's history, beliefs, and traditions, making heritage textiles unique.

Examples of culturally significant handwoven fabrics include:

- **Ikat and Sambalpuri Weaving (India):** These handwoven textiles involve intricate resist-dyeing techniques that create blurred yet vibrant patterns, preserving centuries-old craftsmanship.
- **Handwoven Textiles in Southeast Asia:** In regions such as Bali, Nusa Tenggara Timur (NTT), and Toraja, weaving is a sacred practice, with textiles often used in ceremonial and religious contexts [4].

2.1.3 Economic and Social Empowerment

Handloom industries contribute significantly to the economic well-being of rural and indigenous communities. Weaving provides:

- **Sustainable Livelihoods:** Many artisans, especially women, rely on handweaving for income, helping sustain traditional knowledge and local economies.
- **Fair Trade and Intellectual Property Protection:** Government initiatives and fair-trade policies ensure that artisans receive fair compensation while safeguarding their craft against mass production.

2.1.4 Fusion of Tradition and Modernity

In recent years, traditional handweaving techniques have been revived through collaborations with contemporary fashion designers. Heritage fashion bridges the gap between tradition and modernity and holds relevance in the global marketplace. For example, artisans are now testing color combinations and using undiscovered fabric combinations, such as Eri silk, cotton, and wool, to increase functionality and appeal to the market.

Heritage fashion's traditional crafts, particularly through handweaving, represent a sustainable, ethical, and culturally engaged way of making textiles. Eco-friendly and ethical fashion is on the rise, which makes the promotion and preservation of handweaving practices essential to sustaining craft and skill and creating a more ethical fashion system.

2.2 *Embroidery and Its Cultural Significance*

Embroidery is an ancient textile art form embedded in cultural heritages in many areas of the world. It is a relief technique as well as a means of storytelling, representing various communities' histories, social experiences, and religious histories. The varied embroidery approaches have been maintained and conveyed through many generations and continue to be a part of the cultural identity of many communities.

2.2.1 Cultural and Artistic Value of Embroidery

Embroidery holds an important role in heritage fashion, reflecting the differing social, economic, and spiritual values of varying cultures. For example, traditional embroidered garments, Phulkari (flower work), from Punjab, India, are understood to symbolize prosperity and good fortune. Phulkari is often gifted during weddings and significant occasions. The Hmong embroidery of Thailand is another example of added decoration, along with being a way to transmit culture. Traditionally, the Hmong people incorporated symbolic messages into their embroidery, often using motifs that illustrate geometric patterns and everyday life. Miao embroidery from China is an additional example of cultural storytelling through fabric. The Miao craftsmanship of fabric embroidery has been labeled an intangible cultural heritage. Miao embroidery designs in their craft represent myth, migration history, and spiritual beliefs [5]. The craftsmanship has developed a vibrant craft form in contemporary fashion while continuing to reflect symbolic cultural expressions.

2.2.2 Regional Embroidery Styles and Techniques

Specific regions have their embroidery styles, each distinct in technique, motifs, and cultural background:

- Phulkari (Punjab, India): Uses floral and geometric designs to create intricate embroidery using a darning stitch on coarse cotton fabric. Traditionally, Phulkari has been done by women and holds a significant cultural identity for Punjabis.
- Hmong Embroidery (Thailand, Vietnam, Laos, and China): Involves intricate appliqué and reverse appliqué techniques to create the design, where layers of fabric are stitched together and cut to create the design. The motifs feature nature, prosperity, and spirituality.
- Miao Embroidery (China): Known for brilliant color schemes and geometric patterns while it is typically done with silk threads that have metallic features. Miao embroidery is also used to embellish traditional clothing and accessories.

- Fergana Valley Embroidery (Central Asia—Uzbekistan, Tajikistan, and Uighur communities): Traditional embroidery includes suzani, dorpech, and more that includes flowers and animal motifs. The motifs symbolize nature, fertility, and cultural heritage and continue to be widely applied to decorative textiles [6].

2.2.3 Sustainability and Revival in Contemporary Fashion

Embroidery techniques, especially those relying on handcraftsmanship, support sustainable fashion practices by promoting slow production and ethical artisan work. Heritage embroidery techniques have been adapted by modern designers to preserve traditional crafts while making them relevant in contemporary markets.

The revival of handwoven textiles within luxury fashion emphasizes the persistent relevance of traditional embroidery in fashion across the globe. Traditional embroidery and craft provide an enduring relevance to the future of fashion as it merges ancient craft into contemporary design while forging a link between culture and fashion.

2.3 *Natural Dyeing and Eco-Friendly Alternatives*

Natural dyeing is a historic practice that has been an integral part of textile culture around the world. As an increasing number of fashion personalities and brands become aware of sustainability, natural dyeing has again gained attention because of its sustainability, non-toxicity, and support of local artisans. Unlike synthetic dyes, which contribute to environmental pollution, natural dyes are derived from renewable sources, such as plants, minerals, and insects [7].

2.3.1 Historical Significance of Natural Dyeing

The use of natural dyes dates to the Bronze Age, where early civilizations used plant extracts, crushed minerals, and even insects to color fabrics. Some of the most well-known traditional natural dyes include:

- **Madder (*Rubia tinctorum*)**: A red dye extracted from plant roots, used extensively in Persian and Indian textiles.
- **Indigo (*Indigofera tinctoria*)**: A blue dye obtained from plant leaves, widely used in African, Indian, and Japanese textiles.
- **Turmeric (*Curcuma longa*)**: A bright yellow dye sourced from turmeric roots, commonly used in South Asia.

2.3.2 Eco-Friendly Benefits of Natural Dyeing

Natural dyes are increasingly being recognized for their environmental advantages, particularly in contrast to synthetic dyes, which are often toxic and non-biodegradable. The benefits of natural dyes include:

1. **Biodegradability:** Unlike synthetic dyes, natural dyes do not leave harmful residues and can be safely disposed of.
2. **Non-toxic and Non-carcinogenic:** Many synthetic dyes contain heavy metals and azo compounds, which are linked to health risks. Natural dyes, however, do not pose such threats.
3. **Lower Water Pollution:** The production and use of synthetic dyes result in significant water contamination due to chemical runoff, whereas natural dyeing processes use fewer harmful chemicals.
4. **Sustainable Sourcing:** Many natural dyes come from agricultural by-products, reducing waste and supporting local farmers.

2.3.3 Mordants in Natural Dyeing

Mordants are substances that help bind dyes to fabrics, enhancing colorfastness. Traditionally, natural mordants such as tannins from tree bark, alum (potassium aluminum sulfate), and iron salts have been used. However, modern eco-friendly alternatives avoid toxic mordants like chrome, copper, and tin, which can be harmful to the environment.

2.3.4 Revival of Natural Dyeing in Contemporary Fashion

With increasing demand for sustainable fashion, many designers and textile artisans are reviving traditional natural dyeing techniques. Several fashion brands are now incorporating hand-dyed fabrics into their collections, blending heritage craftsmanship with modern aesthetics. Innovations such as plant-based indigo vats and organic dyeing cooperatives have helped scale up natural dyeing while maintaining ethical production standards.

Natural dyeing represents an essential part of heritage fashion, offering a sustainable and culturally rich alternative to synthetic dyeing methods. As eco-conscious fashion gains traction, natural dyes will continue to play a crucial role in promoting sustainable textile practices while preserving traditional craftsmanship [8].

3 Heritage Fosters Sustainable Fashion: Principles of Slow Fashion

The concept of slow fashion has emerged as a response to the environmental and social challenges posed by fast fashion. Rooted in heritage and traditional craftsmanship, slow fashion emphasizes sustainable production, ethical labor practices, and conscious consumption. This chapter explores the principles of slow fashion and how heritage fosters sustainable fashion [9].

3.1 The Principles of Slow Fashion

Slow fashion operates on a set of principles that prioritize quality, environmental consciousness, and ethical production. These principles ensure a sustainable alternative to the harmful fast-fashion model [10].

3.1.1 High-Quality and Durable Materials

- Slow fashion embraces premium textiles like organic cotton, linen, and wool for building longevity.
- While fast fashion relies on synthetic fibers, leading to microplastics and synthetic waste, slow fashion focuses on biodegradable and recyclable materials.

3.1.2 Ethical Production and Fair Wages

- Slow fashion advocates for fair wages, safe work environments, and ethical labor standards, typically already working with local artisans and small producers.
- Heritage-based brands collaborate with indigenous artisans to facilitate traditional craftsmanship while also contributing fair employment opportunities.

3.1.3 Minimal Waste and Zero-Waste Design

- Slow fashion companies utilize zero-waste design approaches to minimize fabric wastage and other wasted finished fabric during production.
- The strategies include pattern optimization, upcycling, and repurposing materials to minimize environmental impacts.

3.1.4 Local and Sustainable Sourcing

- Sourcing materials locally reduces the carbon footprint associated with transportation.
- Heritage textile techniques, such as handweaving and plant-based dyeing, minimize energy use and chemical pollution.

3.1.5 Timeless and Versatile Designs

- Slow fashion rejects seasonal trends; it is concerned with timeless, adaptable clothing designs meant to last for multiple years.
- Functional and adjustable clothing designs inspired by heritage designs also embody and extend wearability over time.

3.1.6 Consumer Awareness and Mindful Consumption

- Informing consumers about the influence of fast fashion allows them to develop better consumer habits.
- Slow fashion brands promote repairing, recycling, and upcycling their clothing instead of throwing it away.

3.2 Heritage's Role in Fostering Sustainable Fashion

Heritage is essential to sustaining slow fashion because it helps to maintain the cultural transmission of craft, and in particular, the codification of techniques that are often inherently sustainable.

3.2.1 Preservation of Traditional Craftsmanship

- Artisanal handweaving and embroidery techniques, as well as natural dyeing processes, are passed down from generation to generation to support sustainable production.
- By supporting artisans, we are sustaining the local economy as well as avoiding the loss of cultural heritage.

3.2.2 Eco-Friendly Indigenous Textiles

- Indigenous textile manufacturing, like khadi textiles in India and batik textiles in Indonesia, generally has low energy and water requirements.
- They are typically made from organic fibers and colored with natural dyes; therefore, they are biodegradable and non-toxic.

3.2.3 Small-Scale and Slow Production

- In contrast to fast fashion, which results in mass production, heritage fashion promotes the use of craft-based small production methods, reducing overproduction and the waste stream.
- Slower production cycles result in higher-quality garments with less environmental impact.

Slow fashion, rooted in heritage, provides a sustainable alternative to the fast-fashion industry. By prioritizing ethical labor, durable materials, and mindful consumption, slow fashion promotes a responsible and environmentally friendly approach to clothing. Heritage-based techniques and cultural craftsmanship play a crucial role in sustaining this movement, ensuring that fashion remains a force for positive social and environmental change [11].

4 The Impact of Fast Fashion on Sustainability

Fashion has long been an expression of culture, history, and identity, with heritage playing a crucial role in shaping sustainable practices. However, the rise of fast fashion has disrupted traditional textile craftsmanship, leading to significant environmental and social consequences. The role of heritage in fostering sustainable fashion and the detrimental impact of fast fashion on sustainability are discussed further.

4.1 The Significance of Heritage in Sustainable Fashion

Heritage fashion encompasses traditional textile techniques, artisanal craftsmanship, and locally sourced materials that have been passed down through generations. These methods prioritize durability, ethical labor practices, and minimal environmental impact. Sustainable fashion brands and initiatives increasingly draw inspiration from heritage practices, integrating them with modern technology to promote eco-conscious production.

Key Aspects of Heritage-Based Sustainable Fashion

- **Artisanal Craftsmanship:** Handwoven fabrics, natural dyes, and slow production methods reduce waste and energy consumption.
- **Local Sourcing:** Traditional materials such as organic cotton, wool, and plant-based dyes minimize carbon footprints.
- **Longevity and Timelessness:** Unlike fast fashion's ephemeral trends, heritage garments are designed for longevity, reducing overall textile waste.
- **Cultural Preservation:** Supporting local artisans helps sustain traditional crafts and provides economic opportunities for indigenous communities.

4.1.1 The Fast-Fashion Industry and Its Environmental Impact

Fast fashion, characterized by rapid production cycles, low-cost materials, and frequent trend turnover, has severe environmental and social implications. The industry's unsustainable practices contribute to pollution, excessive resource consumption, and exploitative labor conditions.

Environmental Consequences

1. **Textile Waste and Landfill Overflow**
   (a) Fast fashion encourages a disposable clothing culture, leading to excessive waste.
   (b) The average consumer discards 80 pounds of clothing per year.
2. **Water Pollution and Overuse**
   (a) Textile dyeing is a major source of industrial water pollution.
   (b) A large quantity of water is required for cotton production, draining natural resources.
3. **Synthetic Fibers and Microplastic Pollution**
   (a) Polyester, a widespread textile in fast fashion, is known to shed microplastics into bodies of water, adversely affecting marine ecosystems.
4. **Carbon Footprint**
   (a) The dependency on fossil fuel-based fibers and global supply chains in fast-fashion production has a serious impact on the production of greenhouse gas emissions.

Social and Ethical Issues

1. **Labor Exploitation**
   (a) Workers in the garment industry in developing countries endure hazardous conditions and low pay.
   (b) The 2013 Rana Plaza disaster highlighted the dire consequences of neglecting worker safety.

2. **Consumer Misinformation**
   (a) Greenwashing misleads consumers about sustainability efforts in fast-fashion brands.
3. **Cultural Erosion**
   (a) The mass production of synthetic, cheaply made apparel marginalizes traditional artisans and weakens cultural identity.

4.1.2 Bridging Heritage and Sustainable Fashion

To counter the detrimental effects of fast fashion, the industry must integrate heritage-based principles into mainstream sustainability efforts.

Potential Solutions

- Adoption of Slow-Fashion Practices: Encouraging consumers to invest in quality, timeless pieces rather than disposable trends.
- Innovation in Sustainable Fibers: Utilizing biodegradable and recycled materials, such as Lyocell and organic cotton.
- Consumer Education and Advocacy: Raising awareness about ethical consumption and promoting responsible purchasing habits.
- Corporate Accountability: Holding fashion brands accountable for their supply chain transparency and fair labor practices.

Heritage fashion provides a viable path toward sustainability by prioritizing ethical craftsmanship, environmental stewardship, and cultural preservation. Fast fashion's harmful impact on sustainability underscores the urgent need for a paradigm shift in the industry. By integrating traditional practices with modern sustainable innovations, the fashion industry can create a more responsible and lasting future.

4.2 Heritage Fosters Sustainable Fashion: High-Quality Craftsmanship and Timeless Designs

Heritage plays a vital role in fostering sustainable fashion through high-quality craftsmanship and timeless designs. Traditional techniques, passed down through generations, emphasize durability, ethical production, and cultural significance. This chapter explores how craftsmanship and timeless design principles contribute to sustainable fashion and counteract the environmental and social damage caused by fast fashion.

4.2.1 The Value of High-Quality Craftsmanship in Sustainable Fashion

Craftsmanship is at the heart of sustainable fashion, as it prioritizes meticulous handwork, ethical labor, and superior materials that ensure longevity [12].

1. **Skilled Artisan Techniques**
   - Handwoven textiles, embroidery, and natural dyeing methods reduce reliance on industrial production and toxic chemicals.
   - Luxury craftsmanship is deeply connected to tradition, maintaining heritage while ensuring sustainability.
2. **Ethical and Local Production**
   - Supporting local artisans helps sustain traditional skills and provides fair wages.
   - Heritage-based brands often collaborate with small-scale producers, promoting ethical working conditions.
3. **Durability and Quality over Mass Production**
   - Heritage craftsmanship ensures garments are made to last, reducing waste from disposable fashion.
   - High-quality materials such as organic wool, linen, and silk are favored for their durability and biodegradability.

4.2.2 Timeless Designs: A Sustainable Approach

Timeless fashion rejects fleeting trends in favor of classic silhouettes, multi-seasonal wear, and adaptability.

1. **Classic and Versatile Silhouettes**
   - Timeless designs remain relevant across seasons, reducing the need for constant wardrobe turnover.
   - Minimalist cuts and neutral colors extend the life cycle of garments.
2. **Adaptable and Multi-Functional Garments**
   - Transformable clothing, inspired by heritage patterns, allows wearers to style garments in multiple ways.
   - Adjustable and one-size-fits-all designs accommodate diverse body shapes and reduce textile waste.
3. **Emotional and Cultural Connection**
   - Heritage or ethical fashion promotes emotional durability, as clothes can hold a cultural, sentimental, or historical value for individuals.
   - If clothes have historical significance, consumers are more likely to repair or maintain those items.

Skilled craftsmanship and timeless design values are cornerstones of sustainable fashion. The preservation of cultural or craft-based ethical labor, as well as an emphasis on durability, allows heritage fashion to be a responsibly practicable alternative to fast fashion. By offering enduring aesthetics, heritage fashion directly correlates with prolonged wear periods for garments, which minimizes waste and promotes conscious consumption.

5 Collaboration Between Artisans and Designers

Collaboration between artisans and designers is a key aspect of sustainable fashion. By bringing together traditional skills with cutting-edge design, collaborations can yield fashion products that support cultural preservation, ethical production, and sustainability. This section will examine the value of artisan/designer collaborations, how they promote sustainable fashion, and the challenges that they encounter.

5.1 The Value of Artisan–Designer Collaborations

Collaborations between artisans and designers create a synergy that brings together both conventional and modern fashion through the integration of heritage and contemporary aesthetics.

5.1.1 Enhancing Sustainability Through Traditional Craftsmanship

- Crafts that are considered “traditional” support sustainability efforts as they use “natural” materials, develop hues through natural dyes, and practice techniques that result in zero waste of materials.
- Craft artisans possess such deep knowledge regarding the materials they work with that emerging approaches to fashion product development (i.e., product longevity, quality, etc.) can be realized [13].

5.1.2 Preserving Cultural Heritage

- Working together supports the retention of indigenous practices and the preservation of traditional skills lost through industrialization.
- The incorporation of heritage craftsmanship in modern fashion cultivates an appreciation of culture and authenticity.

5.1.3 Ethical and Fair-Trade Practices

- These partnerships support local artisan communities with more equitable pay, leading to productive and sustainable livelihoods.
- Co-design models ensure artisans receive recognition and equitable participation in the creative process [14].

5.2 Challenges in Artisan–Designer Collaborations

Despite their benefits, these partnerships face several challenges that need to be addressed for long-term success [15].

5.2.1 Bridging Design and Craftsmanship

Designers may struggle to fully understand artisans' traditional techniques, leading to misinterpretations of craft value.

- Artisans may find it difficult to adapt to modern design trends while maintaining their traditional methods.

5.2.2 Communication Barriers

- Differences in language, work processes, and creative perspectives can hinder collaboration.
- Lack of mutual understanding can result in conflicts over design ownership and creative direction.

5.2.3 Market Accessibility

- Small-scale artisan collaborations often struggle with marketing and distribution, limiting their reach to global markets.
- High production costs associated with handcrafted fashion can make it difficult to compete with fast-fashion brands.

Collaboration between artisans and designers is essential for fostering sustainable fashion by preserving heritage, promoting ethical labor, and producing high-quality, eco-friendly garments. Overcoming communication and market challenges can further enhance these partnerships, ensuring a sustainable and culturally rich future for the fashion industry [16].

6 Circular Fashion and Waste Reduction

6.1 Upcycling and Repurposing in Heritage Fashion

Upcycling and repurposing in heritage fashion transcend mere material reuse; they function as powerful tools for weaving new narratives from the threads of the past. This section explores how these practices serve as a conduit for preserving cultural heritage, fostering sustainability, and promoting a deeper connection to the stories embedded within garments.

Heritage fashion inherently holds history. Within this context, upcycling becomes a sort of material memory, one that allows the designer to encapsulate the physicality of the past into new creations. The newly formed or repurposed garment becomes a palimpsest, capturing the remnants of its previous life, but still maintains a contemporary identity. The act of recontextualizing allows history to be reinterpreted and valued in the present day.

Upcycling in heritage fashion is an element of circular and sustainable fashion systems; it mitigates the issue of textile waste and contributes to more responsible consumption. Sustainable fashion design moves away from the linear nature of consumption and responds instead to a more restorative fashion mode. Upcycling promotes the longevity of existing garments and materials and mitigates the harm of textile production and consumption to the environment. This is particularly relevant in heritage fashion: using high-quality, often irreplaceable materials means it is even more important to recycle and preserve them [17].

Heritage fashion, by its definition, refers to fashion with a history to it. Upcycling, in this context, could be seen as "material memory," whereby the designer is enabling the new piece to contain the physical essence of the original. The upcycled piece gains life as a palimpsest of past gestures, such as stitching, a shape, or a pattern, but at the same time, it becomes a new artifact, made in the contemporary moment. The upcycling process offers a way to revisit and reframe history for the present day.

Upcycling also deepens the connection with craft and the craft processes and techniques of the past. As a side note, traditional craft techniques are a rich resource for sustainable innovation because they reimagine the invention of textiles. Utilizing eclectic decorative techniques revives traditional approaches to garment-making and preserves cultural heritage through the reimagining of techniques, while enabling the construction of new, more sustainable ways to produce fashion. The handwork, visible stitching, or embellishment all highlight the decentered labor of care, conspicuous labor, as well as a respect for the material [18].

Additionally, upcycling in heritage fashion enables the creation of distinctive, unique articles and a sense of individuality and authenticity. The modern consumer's desires for products that reflect personal values and identities are increasing. Repurposed heritage pieces offer a unique option with an inherent narrative history attached to the item and an unusual aesthetic that replaces mass-produced garments, offering identity and meaning more than a product. The narrative component of an upcycled heritage garment does not stop at the material itself but belongs to the artisans who built the pieces, the people who wore them, and the location and time that they were produced in Armstrong et al. [19].

More than passing an upcycled garment down from one generation to another, if designers take the time to share this narrative and have it lived on in the garment, they can give the wearer an even greater appreciation for the value of clothing and could shift future behaviors about the consumer's clothing. More than just rearranging materials, upcycling and repurposing are changing stories: weaving new stories from the threads of the past. They create items that can be beautiful and sustainable while also being culturally relevant and emotionally resonant.

6.2 *Role of Heritage Brands in Reducing Textile Waste*

Heritage brands, with their long histories and significant resources, have a special opportunity to tackle the issue of textile waste. They are not simply manufacturers of products; they are the custodians of textiles, and they can enact change in the practices of the industry and the behaviors of consumers, toward a more sustainable way of thinking. The values of heritage brands are rooted in longevity, and they regard sustainability practices as inherently more worthwhile than immediate gains. In the realm of sustainable luxury, for example, heritage brands are often positioned to do the right thing and to honor ethical and environmental principles. The long game allows heritage brands to devise comprehensive waste reduction strategies that take a long game approach that goes beyond a product cycle's worth of value.

Revitalizing archival materials is one of the most meaningful ways that heritage brands may reduce textile waste. Archival materials are an often-overlooked source of textiles and textile designs that can be reworked and included back in collections as needed. A circular design approach could leverage these archives while lessening the need for virgin material, thus reducing a brand's overall environmental impact. Revitalizing archival materials also performs a symbolic role of demonstrating heritage, and it acts as a conduit from the past to the present [20].

In addition, heritage brands can influence their supply chains and encourage their partners to consider sustainable practices. A brand can leverage its influence to create systemic change within the textile industry by defining a strong set of environmental standards and prioritizing suppliers that meet those standards. This can be particularly impactful in the adoption of recycled and upcycled materials, as well as closed-loop production methods.

Heritage brands have an additional vehicle to influence consumer behavior and sustainability through education on the environmental impact of textile waste and the value of responsible consumption. Brands can orient their audience toward circularity through digital communications. By using their marketing and communication channels (like social media) to foster awareness of the benefits or the need for upcycled, repaired, or recycled clothing, a brand can influence customer consumption practices. Care and repair of textiles, simply extending useful life, or resale/donation of clothing items, prolongs clothing's life.

One of the other important initiatives for heritage brands is establishing repair and refurbishment programs. By introducing an in-house repair service to fix damaged garments, the brand can help to prolong the life of its items and decrease the need for replacements. This reduces waste and reinforces the value placed on craftsmanship and care, which aligns with slow fashion principles that emphasize longevity and durability rather than disposability.

Finally, heritage brands can support policy changes that advance a more sustainable textile industry. These changes can include advocating for initiatives that promote extended producer responsibility, sharing incentives related to recycling and upcycling in textiles, and agency regulation on hazardous chemicals used in the production process. By utilizing their influence, heritage brands can assist in the development of a policy framework that enables a true circular fashion system. Even further, heritage brands have an essential role in reducing textile waste. Through moving toward circularity, utilizing their influence, and educating consumers, heritage brands can help ignite change and push the industry toward a more responsible and sustainable future.

7 Heritage Fashion: A Catalyst for Sustainability

7.1 Cultural Exchange and Influence of Diverse Cultural Backgrounds on Fashion

Heritage fashion acts as an important vehicle for cultural exchange and sustainability, embedding traditional skills and craftsmanship in a contemporary context. As an opposing force to the increasingly dominant and worrying fast-fashion industry, heritage fashion offers an ethical approach to fashion that engages with the preservation of craft and a dialogue around cultural diversity. This study evaluates the role of heritage fashion in the pursuit of sustainability, its potential for cultural exchange globally, and how the international backgrounds of fashion designers are being brought into contemporary fashion.

Sustainability in fashion relates to environmental, economic, and social aspects. Heritage fashion supports sustainability, for example, through the preservation of traditional craftsmanship, as well as reviving and combining traditional textile-making practices that support the continuity of indigenous crafts and reducing mass production reliance. Heritage fashion values quality over quantity, promoting the slow-fashion movement and generating less waste in the clothing and textile industries, along with encouraging sustainable materials. Heritage fashion generates an increasing demand for handmade, culturally immersed garments, partaking in the promotion of artisans, small-scale economies, and localism.

Fashion has always been a reflection of a cultural exchange and has been shaped by globalization, migration, and historical connections. Heritage fashion is part of that exchange, mixing traditional patterns, textiles, and methods with contemporary aesthetics, which makes it marketable to global audiences—a fusion of the old and the new. This form of fashion is a storytelling device through which history, belief, and tradition can be expressed through the medium of clothing while celebrating cultural exchange. The cultural exchange happens naturally, but through ethical agreements between designers and Indigenous communities, clothing items will honor the culture rather than appropriate or exploit it [21].

Fashion has been shaped by many world cultures over the centuries. In contemporary fashion, silk, embroidery techniques, and silhouettes inspired by Asian kimonos showcase Asia's rich textile heritage. Other world cultures, such as African, whose textiles are recognized globally, are influencing high fashion and streetwear. African textiles are becoming well-known globally due to climatic influences, like Ankara prints, developed for the enjoyment of sustainability in high fashion and streetwear. The techniques of Latin American handmade craftsmanship, which utilize alpaca wool, geometric forms, and handwoven textiles, are part of the sustainable fashion movement as well. European heritage is rooted in luxury fashion culture with textiles like brocade, lace work, and brocade, while the designs of Indigenous North America, like Navajo prints and bead work, are used increasingly in ethical fashion collaborations.

Heritage fashion enhances the industry, but concerns associated with cultural appropriation, commodification, and the mass production of cultural-pattern design raise ethical issues [22]. The responsibility still lies with the designers to work with and involve craftspeople and artisan communities respectfully, ensuring fair payment and acknowledgment of role. It is also important for designers to preserve authenticity in the process of transforming heritage for contemporary markets to preserve cultural integrity. Additionally, consumer knowledge is a key contributor to the appreciation of cultural diversity through responsible consumption [23].

Heritage fashion stands as a bridge between past and present, sustainability and commerce, and tradition and innovation. By promoting ethical production, fostering cultural exchange, and valuing craftsmanship, heritage fashion paves the way for a more sustainable and inclusive fashion industry. Moving forward, stakeholders must ensure responsible practices that honor and preserve the rich tapestry of global fashion heritage [24].

7.2 Innovative Sustainable Business Models and Circular Economy Approaches

Heritage fashion is emerging as a key player in the movement toward sustainability by integrating innovative business models that emphasize circular economy principles. The fashion industry has long been criticized for its resource-intensive and wasteful practices, but heritage fashion offers a counter-narrative by preserving traditional craftsmanship, promoting ethical production, and extending product life cycles. Circular economy approaches within heritage fashion focus on reducing waste, reusing materials, and regenerating textiles to create a more sustainable industry [25].

One crucial model in the circular economy is the life cycle sustainability assessment (LCSA), which looks at the environmental, social, and economic impacts of clothing production in the fashion context [26]. LCSA supports heritage fashion brands in being responsible in material sourcing, cutting waste out of the production process, and designing garments for longevity. This aligns with traditional practices, such as natural dyeing, handweaving, and artisanal embroidery, which rely less on synthetic and chemical processes [27].

Additionally, digital transformation is also important for advancing sustainability within heritage fashion. By utilizing digital tools like blockchain for full supply chain transparency, machine learning (i.e., artificial intelligence) for material efficiency and design development, and virtual simulation to prototype clothing without the use of samples, heritage fashion is balancing cultural heritage with digital advancements toward a more sustainable future. Furthermore, an evolving circular business model is adaptive reuse, where historic textiles and vintage clothing are designed into new products. While supporting cultural heritage, adaptive reuse also helps to minimize textile waste in the form of an increasing supply of vintage garments [28].

In addition, the adaptive reuse of cultural heritage within fashion encompasses more than just materials; it also includes sustainable business models intending to assist economic fortitude. For instance, cooperative ownership models, community production networks, and fair-trade certification enable artisans and local craftsmen to create conditions under which heritage fashion can be economically feasible and culturally conscious. Not only do those models provide social sustainability by creating employment, but they also sustain traditional wisdom that could be extinguished by the speed of the fast-fashion notion.

Cultural heritage also brings to the surface the need for consumer engagement within circularity approaches. By creating engagement and education around heritage fashion, the environmental impacts of choices made by the consumer, and the benefits of slow fashion, advocates are able to promote shifts in consumption patterns. Many heritage fashion brands are implementing rental, resale, and repair services to extend the life and possibility of wearing garments in response to provenance and community ethos, preventing the need for more production and reducing the environmental footprint.

Ultimately, heritage fashion stands at the intersection of sustainability and cultural preservation, offering innovative business models that align with circular economy principles. By embracing adaptive reuse, digital transformation, and ethical production, heritage fashion not only addresses the challenges of sustainability but also ensures the continued relevance of traditional craftsmanship in the modern world. As the industry moves toward a more responsible future, heritage fashion serves as a powerful example of how innovation and tradition can coexist to create a more sustainable and culturally enriched fashion landscape [29].

8 Conclusion

Heritage fashion serves as a powerful catalyst for sustainability, intertwining traditional craftsmanship with ethical and eco-friendly fashion practices. This chapter highlights the significant role heritage fashion plays in preserving cultural identity while promoting environmental responsibility. By focusing on traditional techniques, such as handweaving, embroidery, and natural dyeing, heritage fashion reduces reliance on synthetic materials, minimizes waste, and fosters sustainable consumption.

A significant component of heritage fashion pertains to its alignment with slow-fashion principles that advocate high-quality craftsmanship, timeless designs, and fair labor relationships. In contrast to fast fashion, which results in excessive waste and environmental degradation, heritage fashion promotes slow production and conscious purchasing behavior. By focusing on high-quality, handmade pieces, heritage brands represent an alternative to a parasitic and wasteful industry of tossing away mass-produced garments.

Another core tenet of heritage fashion is the circular economy. While upcycling, repurposing, and textile recycling extend the life cycle of a garment as well as its value, they also reduce textile waste and carbon footprints. Many heritage brands are offering take-back programs, repair services, and rental models, which also coincides with the notion of more sustainable business strategies and extending the use of garments. All of these affordances allow traditional textiles or heritage to still be fashionably used with much less impact on the planet.

Working together with designers and craftsmen is important to preserve heritage fashion and for innovating heritage fashion with craftsmen/craftswomen. These collaborations help to sustain artisans by creating fair economic opportunities and sustainable livelihoods while enabling modern designers to connect heritage references with contemporary aesthetics. However, we need to figure out how to address market access, cultural appropriation, and how to adapt to modern design sensibilities. This is the foundation of ethical and respectful collaborations.

Moreover, consumer knowledge and education are both important to the shift toward sustainable fashion. By learning about the environmental and cultural value of heritage textiles, consumers can choose to support sustainability and craftsmanship. At the same time, heritage brands should promote transparency and tell an ethical story around their product to improve their influence on sustainable fashions.

In conclusion, heritage fashion is more than a continuation of the past; it is a means of shaping a sustainable future. By embracing traditional practices and joining them with contemporary sustainability principles, the fashion industry can lessen its environmental impact while protecting cultural heritage and creating a more ethical and accountable ecosystem. Looking ahead, ongoing innovation and collaboration and the engagement of consumers will be vital in ensuring that heritage fashion remains a positive and proactive ally in the pursuit of a circular, sustainable future.

References

1. Johnson, J. T., Howitt, R., Cajete, G., Berkes, F., Louis, R. P., & Kliskey, A. (2016). Weaving indigenous and sustainability sciences to diversify our methods. *Sustainability Science, 11*, 1–11.
2. Binti Shafie, S., Binti Kamis, A., & Bin Ramli, M. F. (2021). Fashion sustainability: Benefits of using sustainable practices in producing sustainable fashion designs. *International Business Education Journal, 14*(1), 103–111.
3. Tamta, M., & Mahajan, S. (2021). The novel silk fiber: Eri. *International Journal of Home Science, 7*(1), 101–104.
4. Semuel, H., Mangoting, Y., & Hatane, S. E. (2022). The interpretation of quality in the sustainability of Indonesian traditional weaving. *Sustainability, 14*(18), 11344.
5. Shiyang, W., & Kolosnichenko, O. V. (2024). Study of Miao embroidery: Semiotics of patterns and artistic value. *Art and Design, 7*(3), 98–109.
6. Ковтун, O. A. (2020). The art of church embroidery and traditional culture: Common features. *Vestnik slavianskikh kul'tur [Bulletin of Slavic Cultures], 56*, 226–235.
7. Ado, A., Yahaya, H., Kwalli, A. A., & Abdulkadir, R. S. (2014). Dyeing of textiles with eco-friendly natural dyes: A review. *International Journal of Environmental Monitoring and Protection, 1*(5), 76–81.
8. Choudhury, A. K. R. (2018). Eco-friendly dyes and dyeing. *Advanced Materials and Technologies for Environmental, 2*, 145–176.
9. Solino, L. J. S., Teixeira, B., & Dantas, Í. D. M. (2020). The sustainability in fashion: A systematic literature review on slow fashion. *International Journal for Innovation Education and Research, 8*(10), 164–202.
10. Cataldi, C., Dickson, M., & Grover, C. (2017). Slow fashion: Tailoring a strategic approach for sustainability. In *Sustainability in fashion and textiles* (pp. 22–46). Routledge.
11. Sanjeeda, I., & Taiyaba, A. N. (2014). Natural dyes: Their sources and eco-friendly use as textile materials. *Journal of Environmental Research and development, 8*(3A), 683.
12. Antanavičiūtė, A., & Dobilaitė, V. (2015). Principles of slow fashion application in clothing collection creation. *Environmental Research, Engineering and Management, 71*(2), 54–59.
13. Temeltaş, H. (2017). Collaboration and exchange between "craftsman" and "designer": Symbiosis towards product innovation. *The Design Journal, 20*(Suppl. 1), S3713–S3723.
14. Barcellos, L. F., & Broega, A. C. (2018). *Co-design of designers and artisans, valorisation and communication of the partnership-a reflection for sustainable fashion design*. Global Design.
15. Hu, J., Hur, E., & Thomas, B. (2024). Value-creating practices and barriers for collaboration between designers and artisans: A systematic literature review. *International Journal of Fashion Design, Technology and Education, 17*(1), 25–36.
16. Paru, S. P. (2024). Exploring cultural influences in contemporary fashion design: A comparative study. *International Journal of Advances in Engineering Management, 6*, 64–71.
17. Fletcher, K. (2013). *Sustainable fashion and textiles: Design journeys*. Routledge.
18. Henninger, C. E., Alevizou, P. J., & Oates, C. J. (2016). What is sustainable fashion? *Journal of Fashion Marketing and Management: An International Journal, 20*(4), 400–416.
19. Armstrong, C. M., Niinimäki, K., Kujala, S., Karell, E., & Lang, C. (2015). Sustainable product-service systems for clothing: Exploring consumer perceptions of consumption alternatives in Finland. *Journal of Cleaner Production, 97*, 30–39.

20. Joy, A., Sherry, J. F., Jr., Venkatesh, A., Wang, J., & Chan, R. (2012). Fast fashion, sustainability, and the ethical appeal of luxury brands. *Fashion Theory, 16*(3), 273–295.
21. Dickenbrok, C., & Martinez, L. F. (2018). Communicating green fashion across different cultures and geographical regions. *International Review on Public and Nonprofit Marketing, 15*, 127–141.
22. Anyanwu, E. C., & Chiana, C. A. (2022). Socio-cultural influences on fashion consumption behaviour of university students. *ARRUS Journal of Social Sciences and Humanities, 2*(1), 47–65.
23. Yen, H. Y., & Hsu, C. I. (2017). College student perceptions about the incorporation of cultural elements in fashion design. *Fashion and Textiles, 4*, 1–16.
24. Garzillo, C., Wildman, A., Balenciaga, I., & Izulain, A. (2024). A new approach for cultural heritage adaptive reuse: Circular governance. In *Adaptive reuse of cultural Heritage* (pp. 409–429).
25. Bilan, Y. V., Pimonenko, T. V., & Starchenko, L. V. (2020). Sustainable business models for innovation and success: Bibliometric analysis. *E3S Web of Conferences, 159*(1), 04037.
26. Tomasetta, C. (2017). *The life cycle sustainability assessment approach applied to tangible cultural heritage conservation—Developing a support instrument for cultural heritage management within a circular economy and life cycle thinking perspective* [Doctoral thesis, Università degli studi di Pavia].
27. Boons, F., Montalvo, C., Quist, J., & Wagner, M. (2013). Sustainable innovation, business models and economic performance: An overview. *Journal of Cleaner Production, 45*, 1–8.
28. Colombi, C., & D'Itria, E. (2023). Fashion digital transformation: Innovating business models toward circular economy and sustainability. *Sustainability, 15*(6), 4942.
29. Prause, G. (2015). Sustainable business models and structures for industry 4.0. *Journal of Security & Sustainability Issues, 5*(2), 159–169.

Alternative Textile Fibers as a Solution for Improving Sustainable Fashion

Havva Baskan-Bayrak and Hale Karakas

1 Introduction

Fiber is defined as a hair-like and thin material characterized by its exceptionally small diameter relative to its length, and it is sourced from plants, animals, or various other natural and synthetic origins. It is considered as a textile fiber when it exhibits certain specific qualities such as outstanding fineness, excellent flexibility, adequate strength, a minimum required length, and cohesiveness. Textile fibers typically have a length that is 1000–3000 times greater than their diameter or thickness, resulting in an extremely high length-to-diameter ratio, which in some cases can approach infinity, and they are the fundamental components in the production of fashion and textile products. They can be divided into two main groups, which are natural and man-made fibers. Natural fibers are primarily obtained from plant and animal sources, while man-made fibers are produced either from petroleum-based resources or from regenerated from natural materials. Natural textile fibers such as cotton, flax, wool, and jute have been used in clothing production for over 5000 years. Synthetic fibers began to gain interest in the fashion and textile industry after the invention of rayon in 1930s [1–3].

H. Baskan-Bayrak (✉)
Integrated Manufacturing Research and Application Center, Sabancı University, Istanbul, Turkey

Textile Technologies and Design Faculty, Textile Engineering Department, Istanbul Technical University, Istanbul, Turkey
e-mail: havva.baskan@sabanciuniv.edu

H. Karakas
Textile Technologies and Design Faculty, Textile Engineering Department, Istanbul Technical University, Istanbul, Turkey
e-mail: karakashal@itu.edu.tr

S. S. Muthu (ed.), *Cultural Sustainability in Fashion*, Sustainable Textiles: Production, Processing, Manufacturing & Chemistry,
https://doi.org/10.1007/978-3-032-03207-2_5

In ancient civilizations, textile fibers like cotton, linen, and silk played a central role, shaping cultural identities and facilitating the development of global trade networks. During the Middle Ages, wool emerged as the dominant textile, while the Renaissance period saw the advent of mechanized looms, paving the way for industrialization. Furthermore, the twentieth century experienced a revolution in synthetic fibers, the advent of technical textiles, and the emergence of digital printing technologies. In the twenty-first century, the focus shifted to smart textiles, integrating materials with electronics, nanotechnology, and sustainability, thereby expanding the frontiers of innovation. This chronological perspective underscores how textiles have continually evolved in response to changing times, needs, and opportunities, reflecting the persistent influence of human creativity and technological innovation on their production and functionality [4].

As sustainability gains increasing importance, supported by strategies developed by the European Union (EU), the selection of textile fibers in the clothing industry becomes crucial in reducing hazardous environmental impacts. The fiber selection not only influences resource consumption such as water, energy, and chemicals during their cultivation or production but also affects the manufacturing pathways for a specific garment that are largely determined by the type of raw materials used. Moreover, the garment's end-of-life options, including its biodegradability and recyclability are governed by the type of the utilized fiber. In this sense, natural fibers have taken greater interest from textile manufacturers and agricultural fiber producers [5] as it is essential to address a range of environmental challenges including water and land pollution, overuse of chemicals and freshwater resources, consumption of toxic substances, excessive energy usage, biodiversity loss, unsafe working conditions, and animal cruelty during its use [5, 6]. One of the most common approach to mitigate the environmental impacts of textile fibers and garment production aligning with the EU priorities [7] is the adoption of more sustainable materials, such as recycled or organically sourced fibers. Hence, natural fibers such as organic cotton, hemp, and flax are highly valued for their renewability and comparatively low environmental footprint. At the same time, advancements in recycled and bio-based synthetic fibers present promising alternatives that reduce reliance on virgin resources, supporting a more circular and sustainable textile economy. For example, the development of polylactic acid (PLA) fibers, a bio-based synthetic fiber derived from corn starch or sugarcane, can be regarded as a renewable and compostable alternative to petroleum-based materials [8]. In addition, recycled polyester (rPET) also contribute to sustainability by diverting plastic waste from landfills and reducing the demand for virgin fossil-based resources [9]. Moreover, the development of innovative sustainable fibers as alternatives to conventional fiber types has garnered increasing attention in recent years. For instance, regenerated cellulose fibers made from byproducts from the citrus industry and biomass waste products and bio-leather made from soy or mushrooms contribute both to decrease the environmental footprint and support to achieve sustainable textile production [10]. Bamboo is another promising natural sustainable fiber with significant potential for textile and fashion applications, due to its remarkable environmental and human-friendly features that contribute positively to both ecological well-being and consumer health [11]. Lyocell, commercially known as TENCEL, is also obtained from sustainably sourced wood pulp through a closed-loop solvent system and is popular due to its low toxicity and reduced environmental burden [12].

In light of these developments, it is essential to critically examine the environmental impacts associated with conventional textile fibers and evaluate the potential of bio-based and recycled alternatives as more sustainable solutions. Therefore, in this chapter, the environmental impacts of both natural and synthetic textile fibers are discussed in detail by highlighting their functional features and the challenges starting from the manufacturing and end-of-life stages. In response to the growing need for sustainable alternatives, the chapter further explores a range of innovative fiber solutions including bio-based synthetic fibers such as Sorona® (bio-poly(trimethylene terephthalate)), polylactic acid (PLA), and bio-based polyethylene terephthalate (bio-PET); recycled fibers like recycled polyester (rPET) and recycled cotton; and a variety of sustainable natural-origin fibers such as corn fiber, Gallus-Gallus feather fibers, lotus fiber, pineapple leaf fiber, orange fiber, and lyocell. Each of these alternative sustainable textile fibers is examined with a focus on their environmental advantages, material properties, and capacity to contribute to a more circular and responsible textile economy. Supported by ongoing scientific research and industry initiatives, these innovations play a pivotal role in driving the transition toward an environmentally conscious and circular fashion industry.

2 Environmental Impact of Conventional Textile Fibers

Textile fibers are fundamental elements for sustainable fashion, serving not only as the primary components of every garment but also as critical determinants of a product's environmental and social impact across its entire lifecycle [13, 14]. In order to meet the demands of a rapidly evolving and sustainability-focused global world, they must be engineered to offer enhanced functionality, comfort, and the most importantly sustainable [15]. Conventional textiles, however, present a range of pressing challenges, including substantial ecological footprints, inefficient production processes, reliance on finite resources, and limited end-use performance. Additionally, the use of hazardous chemicals, poor labor conditions in supply chains, and low levels of consumer awareness further amplify these issues. Such limitations highlight the urgent need for a paradigm shift toward more innovative, responsible, and circular approaches to textile fiber development, utilization, and end-of-life management [16].

Conventional textile fibers, broadly categorized into natural and synthetic types, each present distinct advantages and limitations in terms of sustainability. As known, natural fibers originate from renewable resources unlike synthetic fibers that depend on nonrenewable petroleum-based materials, contributing to the depletion of natural resources. Moreover, natural fibers are biodegradable, minimizing waste accumulation at the end of their life cycle (EOL) [17]. For instance, cellulose, the main component of plant-based fibers, can be enzymatically hydrolyzed by microorganisms into glucose at the EOL [18]. These attributes make certain natural fibers comparatively more sustainable than synthetic ones due to their reduced environmental impact. However, the production of natural fibers is not completely sustainable due to consumption of toxic chemicals, large volumes of fresh water, energy to extract cellulose, applying pesticides and hormones in wool production, and harming or killing animals like silkworms for fiber harvesting present ethical and environmental challenges in achieving fully sustainable garment production.

2.1 Environmental Impact of Cotton

Cotton, derived from the *Gossypium* plant, is the most extensively utilized and globally significant natural textile fiber [19]. As a seed-based, single-celled, cellulosic material sourced from renewable origins, cotton is fully biodegradable, making it an environmentally favorable choice. Its versatility allows it for broad application across apparel for men, women, and children. Highly valued by consumers, cotton garments are known for their superior moisture absorbency, breathability, and softness, contributing to their great popularity [20]. In order to enhance its functional performance, particularly its strength, durability, wrinkle resistance, and drying efficiency, cotton is often blended with synthetic fibers such as polyester, enabling the production of cotton textiles that meet diverse consumer and industry demands. Despite its numerous advantages, the cultivation and processing of cotton fibers are associated with considerable environmental and social challenges. One of the most critical concerns is its intensive freshwater demand, with estimates indicating that cotton production consumes between 8000 and 22,000 L of water per kilogram of fiber. This substantial water usage places significant pressure on freshwater resources, particularly in water-scarce regions, raising serious concerns about the long-term sustainability of cotton-based textile production. In contrast, synthetic fibers exhibit markedly lower water consumption compared to natural fibers such as cotton, utilizing approximately only 1% of the water required for cotton cultivation. Moreover, their production processes are generally free from harmful chemical applications. For instance, the manufacturing of acrylic fibers demands as little as 0.3×10^{15} L/kg of water, while polyester fibers require just 17.2 L/kg. These quantitative findings underscore the relative water efficiency of synthetic fibers, positioning them as resource-conserving alternatives in terms of water usage. Nevertheless, the energy consumption for producing synthetic fibers is considerably higher with the processes involved in their manufacture such as polymerization, spinning, and finishing. For instance, the energy required to produce polyester fibers ranges from 369 to 432 MJ/kg, substantially exceeding the 38 to 46 MJ/kg required for cotton [21]. This significant difference in energy consumption reveals a key trade-off in evaluating the overall sustainability of natural versus synthetic textile fibers [22, 23].

Cotton cultivation is notably sensitive to a broad spectrum of pests and insects, necessitating the intensive application of pesticides and insecticides throughout the fiber production process. Despite occupying only approximately 3% of the world's agricultural land, cotton cultivation accounts for nearly 11% of global pesticide usage and as much as 26% of insecticide consumption in conventional farming practices [24]. In addition, cotton cultivation involves the treatment of seeds with insecticides, the application of herbicides to suppress weed proliferation, and the use of synthetic fertilizers (approximately 33% of the weight of cotton fiber production) to enhance soil fertility. The extensive use of synthetic agrochemicals in cotton cultivation not only contributes to acute environmental toxicity such as degrading soil quality, diminishing fertility, and contaminating adjacent ecosystems [25] but also poses significant health risks to humans, including carcinogenicity, teratogenic effects, tumor formation, and genetic mutations [26]. In addition to this, workers involved in cotton farming and harvesting may be exposed to these risks due to inadequate protective measures.

Conversely, processes such as scouring and mercerization of cotton fibers contribute significantly to water pollution, as they involve the extensive usage of caustic soda (sodium hydroxide, NaOH). The effluents generated, contaminated with residual alkaline chemicals, are generally discharged into aquatic systems, causing serious environmental concerns. Furthermore, the application of durable press finishes to cotton fabrics often involves the use of formaldehyde, a known carcinogen material, contributing to acute aquatic toxicity upon discharge into water bodies. Its presence also presents substantial health hazards to workers exposed during manufacturing processes [27]. Consequently, traditional cotton fibers still present a challenge in the sustainable textile production due to their intensive resource demands and environmental footprint.

2.2 Environmental Impact of Wool

Wool, a widely used animal-based natural fiber primarily obtained from sheep, is valued for its renewability and biodegradability, making it an environmentally favorable textile material. Wool is highly favored for its versatility in creating garments suitable for both winter and summer, due to its outstanding durability, exceptional moisture absorbency, resilience, fire resistance, and warmth [28]. Wool-based textiles extend their usage beyond apparel, finding application in thermal and acoustic insulation across multiple industries, as well as serving as effective absorbent materials for oil spill remediation [29]. Moreover, EOL wool garments can be efficiently recycled into new products, contributing to waste reduction and preventing their accumulation in landfills [30].

On the other hand, in order to maintain sheep health and prevent parasitic infestations, chemical pesticides and insecticides are commonly applied to the animals [25]. In addition, chemicals applied to prevent felting shrinkage in wool fibers are hazardous. These chemicals remain in the wool fibers and are released during the wool scouring process, ultimately being discharged into water sources causing contamination of water and land, as well. For instance, the utilization organophosphates, toxic elements, during the 1990s resulted in serious health issues among numerous sheep farmers [31]. Beyond this, another environmental concern of wool fiber is the need of an extensive land for sheep farming, which causes deforestation, soil erosion, and decreased biodiversity [32]. This poses a potential threat to soil fertility and necessitates the cultivation of fodder crops, which could occupy up to 50% of the world's total agricultural land. Furthermore, due to their high intake of plant-based feed, sheep produce significant amounts of methane (CH_4)[33], a high-impact greenhouse gas. On average, a single sheep can emit approximately 30 L of CH_4 per day into the atmosphere [31]. Considering these factors, wool fibers cannot be classified as a sustainable fiber source. Therefore, because of the negative environmental impacts such as risks to human health, and the ethical concerns related to animal welfare, it is imperative that designers and fashion brands must focus on alternative sustainable raw materials instead of wool.

2.3 Environmental Impact of Silk

Silk is a crucial natural fiber obtained from the cocoons of silkworms. Silk fabrics have been widely used throughout history and are renowned for their shine, luster, high strength, and durability among natural fiber-based textiles [34]. There has always been a significant demand for specific silk fabric types such as chiffon, georgette, silk satin, and organza, for a variety of clothing applications such as suits, shirts, nightgowns, evening wear, blouses, sport coats, pullovers and dresses [35].

While the environmental impact of silk sericulture is relatively lower, it is not entirely free from concerns regarding animal cruelty. Silk fiber is typically extracted from cocoons through the cruel method of killing the silkworms by boiling them in hot water, although there is an alternative approach which allows the silk moth to emerge and then collecting the silk fiber [36].

Silk sericulture is predominantly practiced in hot climates across Asia, including countries such as India, China, Korea, Vietnam, and Thailand [37]. In these regions, maintaining precise temperature and humidity levels is essential for optimal silkworm growth, which requires significant energy consumption. Moreover, after harvesting, silk farms use hot air or steam to dry the cocoons, a process that is also energy-intensive. The required energy is typically sourced either by burning mulberry wood or from the electric grid. On the other hand, some silk farms use disinfectants, including lime, formaldehyde, or chlorine, on the cocoons to preserve silkworms from diseases although chemical-free alternatives, such as hot-air sterilization, steaming, and sun drying, are also present for this purpose [38]. In certain cases, silk farms apply growth hormones to fasten silk production. In addition, metallic salts have been used to weight silk, making it heavier and more lustrous. However, this practice can lead to the generation of toxic wastewater and contaminate the silk fibers. Additionally, the silk dyeing and printing industries rely heavily on chemicals, contributing to water pollution [39].

As consumer awareness grows, the push toward ethical silk production methods and eco-friendly processing techniques becomes increasingly vital to ensure the silk's place in a sustainable textile future.

2.4 *Environmental Impact of Polyethylene Terephthalate (PET)*

Polyester, chemically referred to as polyethylene terephthalate (PET), is a type of thermoplastic fiber synthesized from two primary raw materials which are ethylene glycol, a type of alcohol, and terephthalic acid, an aromatic carboxylic acid and covers approximately 52% of the total global fiber consumption. PET fibers are known for their durability, resistance to wrinkles, and versatility in various textile applications such as suiting, shirting, dress, active wear, sportswear, and industrial applications. However, they exhibit inferior thermal comfort compared to cotton, primarily due to its low moisture absorbency and limited breathability. Hence, they are often blended with natural fibers such as cotton or viscose to enhance its thermal comfort properties [39].

The resources used to produce PET fiber are nonrenewable due to being derived from petrochemical sources such as oil and natural gas. The production of PET generally has a lower environmental impact in terms of water consumption and the use of toxic chemicals affecting land, water, and air, when compared to natural fibers. However, its manufacturing process is significantly more energy-intensive, requiring approximately twice the energy of cotton production, which contributes to increased greenhouse gas emissions and may amplify global warming. Furthermore, at the end of its life cycle, clothing made from PET fibers is non-biodegradable and may take up to 200 years to decompose, thereby contributing significantly to plastic pollution. Additionally, during washing and wear, PET garments release microfibers into water bodies such as oceans. These microfibers can be ingested by aquatic organisms, entering the food chain and potentially posing risks to the human digestive system. Addressing these challenges through innovations in recycling, alternative feedstocks, and sustainable textile manufacturing is crucial for the environmental footprint of PET fiber [39, 40].

2.5 *Environmental Impact of Polyamide 6.6*

Nylon fiber, particularly nylon 6,6, is synthesized through a condensation polymerization reaction involving a dicarboxylic acid and a diamine, both of which are derived from nonrenewable petroleum-based resources [41]. This synthetic fiber is extensively utilized in the production of various textile products, ranging from apparel such as activewear, swimwear, stockings, and hosiery to industrial applications including carpets, parachutes, and packaging materials. Similar to PET, nylon fabrics are characterized by their high tensile strength, durability, light weight, and smooth texture. In addition, nylon garments are quick-drying, wrinkle-resistant, and maintain their shape well, making them a preferred choice for both performance-oriented and everyday clothing. Introduced in 1938, nylon was the first synthetic fiber to gain widespread popularity among women, primarily due to its superior durability in stockings compared to silk, offering a longer-lasting and more resilient alternative. However, the production of nylon fiber is highly energy-intensive, consuming approximately three times more energy than cotton manufacturing, thereby contributing significantly to global warming [42]. Moreover, the manufacturing process emits nitrous oxide, a hazardous greenhouse gas that is considerably more harmful than CO_2 in terms of ozone layer depletion [43]. Similar to PET, nylon garments release microfibers during washing, which contribute to marine pollution. These microfibers can be ingested by aquatic organisms, entering the food chain and potentially posing risks to human health, particularly to the digestive system [44].

2.6 Environmental Impact of Viscose

Viscose fiber, also known as rayon fiber, is a semi-synthetic or regenerated cellulose fiber derived from wood pulp, typically sourced from trees such as eucalyptus, pine, or beech. Through a chemical process, the natural cellulose is transformed into a fiber that can closely mimic the desirable characteristics of both silk and cotton, including softness, breathability, and drape. Due to its relatively low production cost, viscose fiber serves as an affordable alternative to natural fibers, making it widely used in the textile industry for applications ranging from clothing and home textiles to industrial products [45].

Although viscose fiber is derived from plant-based raw materials, the manufacturing process is highly energy-intensive and involves the use of several chemicals, which can have significant environmental, health, and social impacts. In the production of viscose fiber, firstly wood is cut into chips, and then the wood pulp is chemically treated, typically with carbon disulfide (CS_2), to dissolve the cellulose. Once the fibers are formed, any excess CS_2 is often released into water bodies, contributing to water pollution. The use of CS_2 poses severe health risks, including an increased risk of coronary heart disease (CHD) for workers who are directly exposed to it. Additionally, exposure to this chemical can lead to birth defects, skin allergies, and even cancer in communities living near viscose fiber production facilities [46]. This highlights not only the environmental impact but also the serious health concerns associated with the viscose fiber manufacturing process [47].

In addition to CS_2, the production of viscose fiber involves the extensive use of sodium hydroxide (NaOH) during the xanthation process. NaOH is a highly caustic and corrosive substance that can cause severe skin burns, eye damage, and harm if ingested. The handling of this chemical poses significant safety risks to workers involved in the manufacturing process [48]. Furthermore, there have been concerning reports of deforestation, particularly of mature trees, to supply the raw materials for viscose fiber production. This not only contributes to environmental degradation but also exacerbates issues related to biodiversity loss and ecosystem disruption. A more sustainable approach would involve cultivating trees specifically for the production of viscose fibers, with a systematic replanting process to ensure the forests are replenished after harvesting. This method would help mitigate the negative environmental impact and promote long-term sustainability in the viscose industry [39].

2.7 Environmental Impact of Acrylic

Acrylic fibers, like PET and nylon, are synthesized from petroleum-based resources through chemical processes. These fibers are primarily produced through the dry spinning method, where acrylonitrile copolymers containing at least 85% acrylonitrile monomer, along with other chemicals. In the dry spinning process, the polymer is first dissolved in dimethyl formamide (DMF), and the resulting solution is then heated before being extruded through a spinneret to form continuous filaments. After extrusion, the filaments are washed to remove residual chemicals, dried, and cut into staple fibers, which can then be used in various textile applications [49, 50].

Acrylic fiber is often used as a substitute for wool, as it provides similar warmth and insulation. Unlike wool, acrylic garments are resistant to acids and are not sensitive to damage from insects or moths. This makes acrylic fibers particularly useful in the production of a variety of clothing items for men, women, and children, especially in colder climates. Acrylic is commonly used to make blankets, carpets, fleece, and pile fabrics. Due to the lower cost of raw materials, acrylic clothing is more affordable than woolen alternatives. As a result, many countries produce a wide range of acrylic-based products, including winter clothing, jumpers, and blankets, catering to the demand for cost-effective, warm textiles [51].

Despite their versatility and affordability, the production of acrylic fibers involves several chemical steps that have implications for both environmental and human health, requiring careful management of the chemicals and processes involved. For instance, the production of 1 kg of acrylic fiber results in the carbon footprint of approximately 5.5 kg of CO_2 equivalent which is a major greenhouse gas contributing to global warming, as well as 0.013 kg of sulfur dioxide (SO_2), which can lead to the formation of acid rain. Moreover, the manufacturing process involves the use of hazardous substances such as arsenic (As), chromium (Cr), cadmium (Cd), and zinc (Zn), which pose serious health risks, including carcinogenic effects from direct exposure, and can also contaminate water and land when released as waste products, exacerbating environmental pollution and posing long-term ecological risks. In addition to the hazardous by-products, the manufacturing of acrylic fiber also generates nonhazardous waste, including empty cans, paper, plastics, and filter cloth. While these materials do not pose immediate health risks, their disposal can contribute to landfill waste and environmental pollution, requiring proper management and recycling to minimize their impact [52].

By adopting technological innovations, integrating sustainable practices, and fostering greater awareness, the textile industry can mitigate these challenges and open new avenues for the development of more responsible and forward-thinking textile solutions.

3 Recycled and Sustainable Fibers

3.1 Bio-based Synthetic Fibers

Increase in synthetic fiber usage is inevitable; however, the dependence in petroleum-based chemicals results in excessive use of nonrenewable sources. It has become among the greatest challenges to obtain bio-based products from renewable resources. Bio-derived materials and biotechnology have gained outstanding popularity and crucialness due to factors including problems relating to instable price and availability of crude oil as a chemical feedstock, economic and environmental problems, growing interest in recycling of manufactured products, and economic drive for emission reduction after Kyoto protocol. This has ended with companies setting new goals of sustainability such as use of renewable sources, reduction of greenhouse gas emissions, and zero growth in energy use. In pursuit to attain these goals, companies have directed their efforts to produce and commercialize fibers from renewable raw materials. Efforts have been focused on research and development of bio-based materials. Bio-based synthetic fibers possess outstanding advantages including low carbon footprint in synthesis process and also most of these fibers offer excellent biodegradability, though there are bio-based fibers that have not proven to be biodegradable. The concerns for sustainability, environmental protection, and efficient resource usage have made it inevitable to explore ideal bio-based raw materials. The bio-based textile fibers offer a great capability and promise for their use in sustainable clothing and other textile applications [53, 54].

Renewable sources to obtain sustainable polymers are mostly starch, cellulose, and chitin, which are chemically modified. Lignin, cellulose, starch, and plant oils are the common biomass from which bio-based polymers are synthesized via a two-step process. Bio-based fibers spun from synthetic polymers from renewable sources may solve the problems of energy consumption and environmental pollution, while offering economic and environmental-friendly source usage. The raw materials used are mostly agricultural, food processing, forestry, wood processing, and marine wastes and by-products, and they are produced by fermentation of these [54, 55]. Currently, most bio-based polymers are polyesters and polyamides. Bio-based fibers development and usage are being fostered [54, 56]. The mostly used and promising bio-based textile fibers will be presented in this part.

3.1.1 Bio-poly(trimethylene terephthalate) (BioPTT, Sorona®)

Sorona family of poly(trimethylene terephthalate) (PTT, or "3GT") polymers from DuPont Bio-Based Materials business unit is among the most successful ones with its unique properties. Sorona® is a polytrimethyl terephthalate (PTT), which is a biopolymer that contains 37% renewable plant-based ingredients. The bio-based ingredient, Bio-PDO™ (bio-based 1,3 propanediol), is made through a fermentation process that uses glucose as the feedstock, mainly from corn.

Polyester, which is the mostly used and consumed fiber in the world, have different varieties including the mostly used fiber polyethylene terephthalate, called also as PET or 2GT. It is possible to obtain other types of polyester polymer and fiber by reacting 1,3-propanediol instead of reacting ethylene glycol with terephthalic acid to make 3GT. 3GT polymers and fibers offer better elastic recovery and lower modulus compared to PET and PBT. The ranking of both the recovery and shrinkage behavior of these materials is in the order 3GT > 4GT > 2GT.

The handicap of 3GT polymer production have been overcome and more economical methods of 1,3-propanediol (PDO, and also called 3G) have been possible and it could be produced from renewable sources in addition to by biological processes. Continuous polymerization processes have been improved to manufacture good quality fibers from 3GT.

Sorona is DuPont's brand of triexta(polytrimethylene terephthalate) that was named and commercialized in 2000. The fibers exhibit softness in addition to high strength and stiffness. Sorona is a co-polymer of 1,3-propanediol (obtained by fermentation) and petroleum-derived terephthalic acid (TPA) or dimethyl terephthalate (DMT).

Sorona® 3GT linear, crystallizable polymer, having a melting temperature of about 2280 C and a glass transition temperature of about 500 °C, has a unique molecular structure. This results in a zigzag shape and in the tensile behavior of a coil spring, exhibiting unique mechanical characteristic. The Sorona® 3GT fiber can have elastic recovery even after 15–20% applied strain after force is removed. The fiber offers all advantages of polyesters including ease of care, good dimensional stability, and weather resistance. Additionally, its dyeability at 1000 °C with disperse dyes and exhibiting very high color fastness, its stain resistance, and resistance to UV degradation are among the fiber's advantages [53].

With regard to environmental footprint, the Sorona® 3GT polymer provides lower greenhouse gas emission) about 40% lower for bio-PDO compared to petrochemical PDO), better energy efficiency compared to polyethylene terephthalate (PET), easier recycling compared to PET and nylon [53, 57].

3.1.2 Polylactic Acid (PLA)

Synthetic fibers processing with incineration and cremation lead to discharge of a great amount of greenhouse gases into the air. Decomposition of plastics require a long period of time, damaging the environment, the ecosystem, and living organisms. Plastics that can be decomposed by living organisms are bio-degradable; they are disintegrated into CO_2, water and biomass. Polylactic acid (PLA) is a thermoplastic aliphatic polyester obtained from sustainable biomass, 100% renewable sources, mainly from aged plant starch like corn, cassava, sugar beet mash, and sugarcane [6]. PLA fiber is regarded as the most important bio-based and biodegradable material. It has a wide range of applications in food packaging, transportation, textiles, and electronics industries.

In spite of its limited biomedical applications, its usability as a textile fiber has important implications. It is an alternative to polyethylene terephthalate (PET), the mostly used and consumed fiber currently in the world. As PET fibers are obtained from nonrenewable fossil fuel resources and as they are non-biodegradable and not easily recycled, PLA stands out as a promising sustainable fiber [58].

PLA offers the advantage of being a melt-processable thermoplastic fiber from renewable natural sources, providing ease of production and processing. It has good physical and mechanical properties, suitable for use as a textile fiber. The fiber has a smooth surface and has a circular cross-sectional shape, though trilobal and cross-sectional shapes are also possible. Its melting temperature of around 160–170 °C can also adjusted to be as low as 1300 °C and as high as 2200 °C. The fiber can attain crimp, can be used as both filament and spun yarns and its tenacity is higher than natural fibers given they are drawn at appropriate draw ratios. The fiber's moisture regain value is between 0.4% and 0.6%, being much lower than natural fibers and slightly above conventional polyethylene terephthalate PET fibers [58–60].

PLA fibers have good wicking, fast moisture spreading and drying, resulting in good moisture management of the clothing made of this fiber. With regard to physiological comfort, test results showed that the apparel made of PLA proved to have better performance compared to cotton and PET. Additionally, the fiber's good elastic recovery and shape retention, high resiliency, and good appearance after washing render PLA a good fiber for apparel applications. Apart from apparel, PLA fibers can be used in home textiles, nonwovens including hygiene, filtration and separation, medical applications, and rigid packaging.

PLA is the most commercially developed fiber among fibers from renewable feedstock, having reached an annual production volume of 200,000 tons in 2020 [56].

3.1.3 Bio-poly(ethylene terephthalate) (Bio-PET)

Bio-PET is non-biodegradable bio-based fiber. It is among sustainable alternative fibers to conventional PET. Bio-PET is produced from renewable sources such as plant-derived ethylene glycol. Its advantage over conventional PET is its processability and recyclability with a lower carbon footprint. This has made it a sustainable alternative fiber with mechanical properties close to PET [61]. In its production, both starting chemicals such as ethylene glycol (EG) and PTA and/or DMT monomers have to be obtained from biological sources. The aliphatic monomer Bio-EG could be synthetized from the hydrolysis of ethylene oxide, obtained via the oxidization of bio-ethylene, which is obtained from the fermentation of glucose, followed by dehydration. Another route to synthetize EG from biomass is via the use of different types of microorganisms [62, 63]. Any method, including methods that get as high as a maximum of 30% bio-PET, offer a remarkably lower carbon footprint compared with the synthesis of oil-derived plastics. This will be remarkable to explore and go after [62].

3.2 Recycled Fibers

Textile sector uses more volumes of recycled fibers due to the increasing awareness of environmental pollution and damage and need for more sustainable fibers and materials as a result of this awareness coupled with legal requirements. Therefore, technological developments and modern methods accelerate recycling technology, leading to the production of recycled fibers displaying performance comparable to virgin fibers. Recycled polyester and recycled cotton are among the mostly used ones in the textile and apparel sector.

3.2.1 Recycled Polyester Fibers

Plastics have become part of our life, finding place in many end-uses. Polyethylene terephthalate dominates the plastics market. PET fibers have surpassed cotton in total global fiber production and consumption, showing an increasing trend. This trend is also expected to continue according to fiber forecasts for 2030. PET finds major applications as material for fibers; injection molded parts, and most importantly for blow molded bottles. PET, having the molecular structure as shown in Fig. 1, is obtained from terephthalatic acid and ethylene glycol through the polycondensation reaction of the monomers. PET production depends upon the chemical reaction between the monomers, which are produced by esterification reactions.

The environmental impacts of polyethylene terephthalate waste have been emphasized for a long time; however, small amount of PET is recycled, in spite of the harm caused by unrecycled PET left [64, 65]. Recycling plastics reduces the environmental damage by decreasing accumulation in the environment. Plastic recycling reduces fossil fuel consumption, energy and municipal solid waste, and additionally carbon dioxide (CO_2), nitrogen oxide (NOx), and sulfur dioxide (SO_2) emissions are also lowered. The target to recycle the majority of plastic waste rather than to send to landfills has been made more possible with the development of advanced technologies and systems of collecting, sorting and reprocessing of

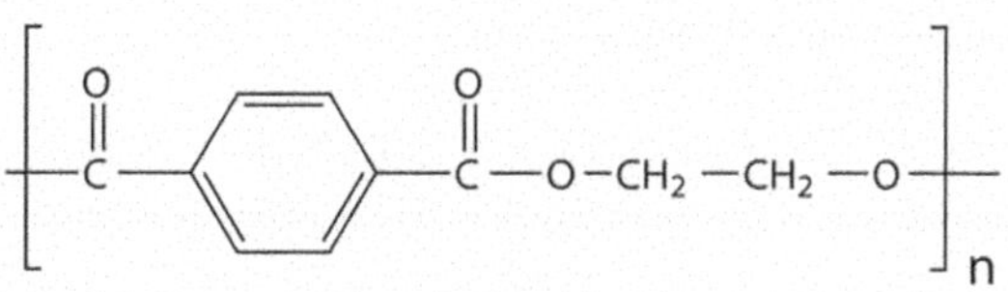

Fig. 1 Poly(ethylene terephthalate) molecular structure

recyclable plastics. Life cycle assessment (LCA) is an important tool to evaluate the uses of recycling programs [66]. Recycling leads to reduction in biological impact, decreasing greenhouse gas (GHG) emissions and fossil fuel consumption [67].

R-PET has many environmental benefits. According to the Environmental Protection Agency (EPA), 2/3 less energy is required to manufacture products made out of recyclable plastic. Other studies show that the production of recycled plastic requires 2/3 less sulfur dioxide, 50% less nitrous oxide, and almost 90% less water usage.

PET recycling methods include mechanical recycling, chemical recycling, and incineration; among these methods studies showed that mechanical recycling was the best method for PET waste management. Mechanical recycling consists of shredding, grinding, melting treatments of PRET waste, and also washing and drying in case of presence of contamination [68].

As virgin PET can be accessed more easily, R-PET has become more expensive in the last years. Recycled PET (R-Pet) fibers, produced from polyester used in bottles, can have various applications, including concrete, nonwoven fabrics, and in yarns to be used in apparel, carpets, packages, industrial goods [64]. The utilization of the recycled material is essential to complete the material cycle. For example, one important end-use is in materials in nonwovens for automotive applications. As nonwovens offer various possibilities to include recycled fiber material (often blended with virgin fibers) in the production of new webs automotive applications, a growing proportion of needle-punched nonwovens for automotive applications is being manufactured from recycled polyester derived from plastic bottles. Furthermore, automotive manufacturers emphasize the use of efficient sound absorptive materials in vehicles to reduce undesired noise. The existing sound absorptive materials are generally made of synthetic fibers, which are not biodegradable [69].

Because of the recyclability of PET and the relative abundance of post-consumer waste in the form of bottles, R-PET is also an important market share as a carpet fiber. After the release of STRAND in 1999, a 100% post-consumer recycled content PET fiber, by Mohawk Industries, more than 17 billion bottles have been recycled into carpet fiber in a period of 8 years [22, 70]. Major supplier to numerous carpet produces, Pharr Yarns supplies a BCF (bulk continuous filament) PET carpet fiber containing a minimum of 25% post-consumer recycled PET content [70, 71].

In a study by Atakan et al.,[72] velour design molded automotive carpets made of recycled polyethylene terephthalate (PET) fibers were developed via needle-punching process to improve their abrasion resistance properties and compared with carpets made of virgin PET fiber. Developed carpets were tested for abrasion resistance performance and they were evaluated according to requested specification. The scanning electron micrograph (SEM) images of R-PET fibers used in the study are shown in Fig. 2.

Results showed that carpets made of 85% rPET +15% bicomponent PET had almost equal performance in terms of both fiber loss and carpet appearances with carpets consisting of 80% PET +20% bicomponent PET. Carpets made of recycled

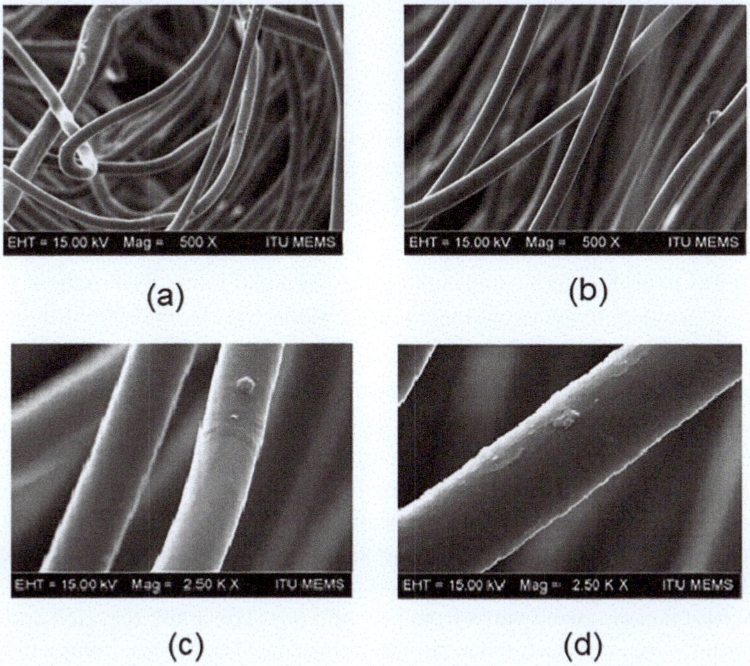

Fig. 2 SEM image of R-PET fibers used in carpets: (**a**) RPET 6 Denier (×250), (**b**) RPET 11 Denier (×250), (**c**) RPET 6 Denier (×1000), and (**d**) RPET 11 Denier (×1000)

PET fibers offer the manufacturer low raw material costs in addition to ecological advantages [72].

Possible end-uses of R-PET are still being thoroughly explored and their performance with its environmental benefits are being evaluated. The fiber has now a well-established state and share in the market. It has been stated that use of additives and also blending with another polymer would be necessary to enhance the properties of the end-uses or to achieve new desirable characteristics. Innovations in the processing methods would also help to increase the industrialization of products made of this fiber [65].

3.2.2 Recycled Cotton Fibers

Cotton is a superior fiber, offering good touch, absorbency, comfort, breathability, and therefore it has always been regarded as a valuable fiber. It finds use in textiles and apparel, its volume of use is second after polyethylene terephthalate (PET) fiber, with its price increasing since 2021 due to increasing global demand of production and consumption [73]. Conventional cotton cultivation is highly polluting the environment due to the use of harmful pesticides and fertilizers. Additionally, it has high water consumption, exploiting the water sources of the earth [74, 75].

Alternative cotton fibers have been proposed in order to decrease the environmental burden and they are promoted versus conventional cotton fibers, recycled cotton being one of them.

Recycled cotton can be generally defined as cotton that is obtained by converting cotton fabric into cotton fiber that can be reused in textile products. Recycled content includes recycled raw material, as well as used, reconditioned, and remanufactured components. Textile recycling is usually generated from two primary sources:

(i) Pre-consumer: includes scraps created by yarn and fabric by-products
(ii) Post-consumer: includes garments, upholstery, towels, household items to be repurposed

Recycled cotton has been sourced from pre-consumer waste mostly as post-consumer waste is more difficult to sort. However, modern technology has made it possible to overcome problems related with different colors, fiber blends, etc.

Recycled cotton fibers obtained from textile waste by mechanical shredding show minor quality results. They are usually blended with other fibers and spun with rotor spinning to be used in coarse yarns with yarn counts between 10 and 20 Ne. These can find usage in denim products, towels, and home furnishing. In the study by Arafat et al., blend yarns from pre- and post-consumer recycled cotton and virgin cotton were manufactured in ring spinning line. The process steps can be seen in Fig. 3. In contrast to the rotor spun yarns with a stiff feeling on the skin, soft ring spun yarns with 30 Ne yarn count, which are also finer, could be achieved. It was concluded that up to 25% recycled cotton fibers can be used as an alternative to virgin cotton to produce medium count (30 Ne) yarn in ring spinning for the production of knit top garments [73].

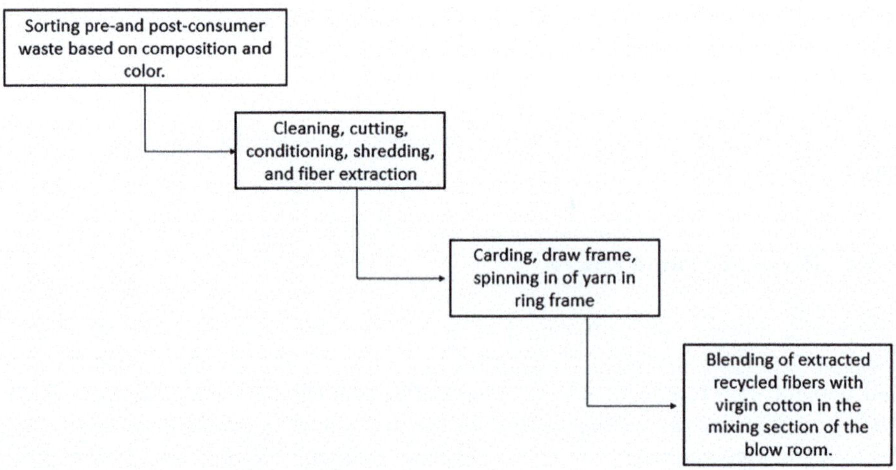

Fig. 3 Flow chart of production of ring spun yarns from blends of virgin cotton and recycled cotton fibers extracted from pre- and post-consumer textile wastes

Recycled cotton yarns, blended with other natural fibers, offer the possibility to be alternative to conventional cotton yarns with their lower environmental impact advantage. Research and development studies in recycled cotton mainly focuses on improving quality problems such as pilling, nep, softness, and also aesthetic of the fabrics made from these fibers.

3.3 Sustainable Fibers from Natural Sources

Ecological and sustainable textile fibers and materials open a platform for creative design and production of these from renewable sources including fruits, vegetables, animal by-products, fishing sector, algae, shellfish, bacteria, plants, and their waste. Their use and reuse are considered for obtaining sustainable fibers from natural sources. There are various nonconventional sources for fiber and fabric production. Banana, orange pineapple, coconut fruits; mil as an animal by-product; fish, shellfish, algae; mycelium from mushrooms and bacteria; plants; and plastics from packaging can be used for the production of ecological fibers and fabrics [76]. Examples for sustainable natural fibers and man-made fibers from natural sources will be presented in this part.

3.3.1 Corn Fiber

Corn, being among the mostly cultivated crops in the world, is a renewable source for fiber and starch. The stalk, husk, and cob of the corn plant are the most important sources for fiber. The cellulose content is the highest in corn fibers in comparison to other sources of fibers like wheat husk, rice husk, wheat straw, wheat bran and rice straw [77]. There are studies and efforts to obtain extracted natural cellulose fibers from cornhusk. One of these studies focused on extraction of fibers from cornhusk by treating with alkali, enzyme, and bleach. The resultant fibers were evaluated for their physical and chemical behavior. The fibers did not prove to have sufficient performance for clothing due to the harsh nature of the fibers; however, they proved to be suitable for potential application in non-clothing technical textile applications [78]. Corn fiber end-uses are mostly in food packaging, medical areas (such as bandages, gauze), wall panels, and baby diapers; there are also clothing applications such as fleece fabrics and denim.

3.3.2 Gallus-Gallus Fibers from Chicken Feathers

Globally, the availability of chicken feathers increases day-by-day due to the consumption of chicken flush in food industry in huge volumes. Feather disposal is 5–10% of the weight of the chicken. The fibers have structural integrity in addition to overall tensile strength, rigidity, and durability, allowing them to withstand mechanical stress and deformation. The fibers have a high moisture regain value of 16–20%. In addition to that, they are lightweight and biodegradable. They are mostly used as fillers, reinforcement materials, insulation materials, and nonwoven fabrics [79–81].

3.3.3 Lotus Fiber

Lotus fiber (*Nelumbo nucifera*) is obtained from the stems of the lotus flower, growing naturally in Burma lakes. The fiber, which is inherently at the micro- to nanofiber range in diameter, has antibacterial, self-cleaning, and superhydrophobic property. The fiber-making process steps are shown in Fig. 4.

Stems of the lotus plants are collected from the lake, they are slightly cut by a knife around the stems, they are twisted, and the exposed fibers are wound on a roller as yarn form. It is a manual process and therefore the speed is very slow. One spinner can produce 250 m of yarn per day. 30 kg of stems give 250 m of yarn. It is a soft and breathable fiber and can be regarded as a perfect example of fibers obtained from stems of plants [82].

3.3.4 Pineapple Leaf Fiber

Pineapple is a tropical fruit, having an important volume of production after banana fruit. Pineapple leaf fiber is obtained from pineapple plant leaves waste (Fig. 5). The pineapple fiber cells are present are as fiber bundles within the leaves of the pineapple.

The fiber has a high cellulose content, high absorption, and good mechanical properties. Table 1 gives a comparison of the properties of pineapple leave fibers with other textile fibers.

It is an ecological fiber as it is biodegradable and recyclable. The fibers are extracted from the leaves after the removal of epidermis layer on the outer surface of the pineapple leaves through biological and mechanical processes, ending with fiber bundles. They are lignocellulosic fibers, containing cellulose, hemicellulose, and lignin besides low amounts of oil, wax, pectin, and ash. They have a whitish color and a soft silk-like touch. Their applications range from clothing including luxury apparel to home textiles, ropes and cordage, technical textiles, conveyor belts, composites, nonwovens, and geotextile applications [83].

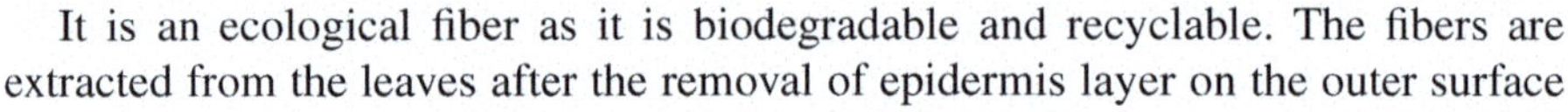

Fig. 4 Process steps in fiber making of lotus fiber

Fig. 5 Pineapple leave fiber threads (*taken from* https://www.ecorivta.com/pineapple-fiber/)

Table 1 Comparison of properties of pineapple leaf fibers with other textile fibers [83]

Fiber type	Density (g/cm^3)	Breaking strength (MPa)	Young modulus (GPa)	Elongation at break (%)	Moisture regain (at standard atmospheric conditions)
Pineapple fiber	1.07–1.52	170–1627	6.21–82.5	1.6	11.8
Cotton	1.5–1.6	287–597	5.5–12.6	3–10	8.0
Flax	1.4–1.5	345–1500	27.6–80	1.2–3.2	10.0
Hemp	1.48	550–900	70	1.6	10.0
Sisal	1.33–1.5	400–700	9–38	2–14	11.0
Jute	1.3–1.46	393–800	10–30	1.5–1.8	12.0

3.3.5 Orange Fiber

One other fiber that is born out of greener alternative solution thinking for more ecological and sustainable textiles is the orange fiber. In order to decrease the impact of the polluting textile industry, grounded orange peel waste was processed so that cellulose could be extracted. Fabrics made from the spun fibers resemble silk as they have a soft touch and good drapability (Fig. 6). The fiber is reported to have microencapsulated vitamins and nourishment in its structure, which can resist up to 20 washing cycles. The fiber uses orange waste instead of sending this waste into landfills. It can be blended with other fibers including cotton, silk, and elastane.

The patented technology of this fiber offers the use of a renewable source of cellulose for textiles and supports an environmentally responsible approach to textile

Fig. 6 Orange fiber fabric (*taken from* https://orangefiber.it/)

manufacturing by reducing pollution and resource consumption. The patented process relies on a chlorine-free chemical treatment, utilizing hydrogen peroxide, which is comparatively eco-friendly producing fewer harmful by-products, safer wastewater, and no toxic residues, as hydrogen peroxide breaks down into water and oxygen. Less water and pesticide usage are also important factors for the environmental responsible character of its production. Raw material dependency for cellulose extraction andsignificant energy usage in chemical treatments can be among major concerns for the production of orange fibers [84, 85].

3.3.6 Lyocell Fibers

Sustainable production of regenerated cellulosic fibers requires cost-effective and environmental-friendly processes. Rayon fibers, which are regenerated cellulosic fibers, have attained a unique position due to their superior performance in textile and apparel applications. Viscose rayon fibers are prone to research and engineering solutions to achieve sustainability.

4 Future Aspects

The main purpose of textiles was to protect people from environmental factors such as rain and cold. However, over time, their role evolved to also consider aesthetic and decorative functions. In the current situation, textiles are increasingly designed to address the growing demands for sustainability, enhanced functionality, smart features, and adaptability, positioning them as key drivers of innovation within the textile and apparel sectors by decreasing the environmental footprint of the fibers.

The development of sustainable fibers, such as bio-based synthetics or fibers produced with minimal chemical use, presents a promising future in the fiber sustainability. Advancements in biotechnology such as the use of microbial processes to produce fibers could reduce the dependence on harmful chemicals and lower the carbon footprint of textile fiber production. These fibers could be part of a more sustainable textile system, where fibers are continuously reused, recycled, and regenerated, helping reduce textile waste. Moreover, research into fiber materials derived from waste, such as agricultural residues, could lead to more sustainable options. Such progresses can contribute to reduce the long-term environmental impact of textile wastes. In addition, improvements in fiber extraction from unconventional feedstocks like pineapple leaves, chicken feathers, or corn husks may streamline production, reduce costs, and enhance competitiveness with conventional textile fibers. Furthermore, the development hybrid textiles, such as the combination of sustainable alternative ones with the conventional textile fibers suggest potential to balance performance, comfort, and sustainability features. On the other hand, innovations in fiber manufacturing technologies can significantly reduce water, energy, and chemical consumption, and hence lower the environmental footprint of textile fiber production. Beyond these, government policies and regulations can motivate sustainable sourcing, the use of renewable resources with sustainability standards, encouraging to lower environmental footprint.

Consequently, promoting sustainable textile production and integrating it into circular economy models will be the key for the sustainable future from the perspective of textile and fashion industry.

5 Conlusions

Textile fibers are the fundamental elements of sustainable fashion that are both shaping aesthetics and performance of garments as well as their environmental and social footprint throughout the entire life cycle. As discussed in this chapter, both natural and synthetic fibers involve complex trade-offs in terms of sustainability.

Natural fibers such as cotton, wool, and silk are biodegradable and derived from renewable sources, yet their cultivation and processing require high water consumption, significant chemical inputs, extensive land exploitation, and raise ethical concerns. On the other hand, synthetic fibers such as PET, polyamides, acrylic, and viscose offer advantages in terms of cost-effectiveness and water efficiency but they are strongly dependent on nonrenewable petrochemical feedstocks, energy-intensive manufacturing, and are persistent in the environment, contributing to microplastic pollution. Therefore, each fiber has its own functional advantages and environmental concerns. Since effective sustainability in textile and fashion industry is of paramount importance, the shift toward alternative, renewable sources has become imperative. Within this scope, alternative sustainable textile fibers have emerged as a transformative solution. Derived from agricultural by-products, fermentation processes, and renewable plant-based feedstocks, bio-based materials like bio-PTT (Sorona®), PLA, and bio-PET illustrate preserving the functional advantages of conventional synthetic fibers by substantially reducing carbon emissions and improving biodegradability. In parallel, the integration of recycled fibers such as R-PET and recycled cotton reflects the growing importance of circular economy principles in extending material lifecycles and reducing overall environmental impact. Furthermore, the investigation of innovative fibers from unconventional natural sources such as corn husk, pineapple leaves, orange peels, chicken feathers, and lotus stems demonstrates the industry's creative potential in valorizing waste and underutilized resources. However, challenges are still present in scaling up, improving performance consistency, and developing infrastructure for sustainable processing and end-of-life management.

Consequently, the enhancements in sustainable textiles can be governed by the synergistic integration of biotechnology, innovative material science, and regenerative system design. By adopting scientific progress with environmental consciousness and social responsibility, bio-based and recycled fibers present a compelling approach toward sustainable future.

References

1. Kozłowski, R. M. (2012). Handbook of Natural Fibres. *Handbook of natural fibres, 1*, 1–620.
2. Sfiligoj, M., Hribernik, S., Stana, K., & Kree, T. (2013). Plant fibres for textile and technical applications. In *Advances in agrophysical research* (pp. 369–398). IntechOpen.
3. Thapliyal, D., Verma, S., Sen, P., Kumar, R., & Thakur, A. (2023). Natural fibers composites: Origin, importance, consumption. *Journal of Composites Science, 7*(506), 1–44.
4. Shah, M. A., Pirzada, B. M., Price, G., Shibiru, A. L., & Qurashi, A. (2022). Applications of nanotechnology in smart textile industry: A critical review. *Journal of Advanced Research, 38*, 55–75.
5. Oliveira Duarte, L., Kohan, L., Pinheiro, L., Fonseca Filho, H., & Baruque-Ramos, J. (2019). Textile natural fibers production regarding the agroforestry approach. *SN Applied Sciences, 1*(8), 1–10.
6. Nayak, R. (2020). *Supply chain management and logistics in the global fashion sector: The sustainability challenge*. Routledge.

7. European Comission. (2021). *EU strategy for sustainable textiles*. European Comission.
8. Trivedi, A. K., Gupta, M. K., & Singh, H. (2023). PLA based biocomposites for sustainable products: A review. *Advanced Industrial and Engineering Polymer Research, 6*(4), 382–395.
9. Periyasamy, A. P., & Militky, J. (2020). LCA (Life Cycle Assessment) on recycled polyester. In *Environmental footprints of recycled polyester* (pp. 1–30). Springer.
10. Sigaard, A. S., & Laitala, K. (2023). Natural and sustainable? Consumers' textile fiber preferences. *Fibers, 11*(2), 1–23.
11. Ali, M. A., & Sarwar, M. I. (2010). *Sustainable and environmental friendly fibers in textile fashion*. University of Boras.
12. Lenzing Group. (2020). *Up for future generations highlights 2020*. Lenzing Group.
13. Fouda, A., Hamza, M. F., Shaheen, T. I., & Wei, Y. (2022). Editorial: Nanotechnology and smart textiles: Sustainable developments of applications. *Frontiers in Bioengineering and Biotechnology, 10*(August), 1–3.
14. Farooq, A. S., & Zhang, P. (2022). A comprehensive review on the prospects of next-generation wearable electronics for individualized health monitoring, assistive robotics, and communication. *Sensors and Actuators A: Physical, 344*(April), 113715.
15. Persson, N. K., Martinez, J. G., Zhong, Y., Maziz, A., & Jager, E. W. H. (2018). Actuating textiles: Next generation of smart textiles. *Advanced Materials Technologies, **3***(10), 1700397.
16. Gonzalez, V., Lou, X., & Chi, T. (2023). Evaluating environmental impact of natural and synthetic fibers: A life cycle assessment approach. *Sustainability (Switzerland), 15*(9), 7670.
17. Muthu, S. S., Li, Y., Hu, J. Y., & Mok, P. Y. (2012). Quantification of environmental impact and ecological sustainability for textile fibres. *Ecological Indicators, 13*(1), 66–74.
18. Muthu, S. S. (2014). *Roadmap to sustainable textiles and clothing: Eco-friendly raw materials, technologies, and processing methods*. Springer.
19. Elfaleh, I., Abbassi, F., Habibi, M., Ahmad, F., Guedri, M., Nasri, M., & Garnier, C. (2023). A comprehensive review of natural fibers and their composites: An eco-friendly alternative to conventional materials. *Results in Engineering, 19*(June), 101271.
20. Ravandi, S. A. H., & Valizadeh, M. (2011). Properties of fibers and fabrics that contribute to human comfort. In *Improving Comfort in Clothing*. Elsevier.
21. Cupit, M. J. (1996). *Opportunities and barriers to textile recycling*. AEA Technology, Recycling Advisory Unit.
22. Muthu, S. S. (2014). *Assessing the environmental impact of textiles and the clothing supply chain*. Woodhead Publishing.
23. Yu, Z., & Yang, Y. (2025). Carbon footprint of global cotton production. *Resources, Environment and Sustainability, 20*(March), 100214.
24. Choudhury, A. K. R. (2017). Green chemistry and textile industry. *Journal of Textile Engineering & Fashion Technology, 2*(3), 351–361.
25. Fletcher, K. (2012). *Sustainable fashion and textiles: Design journeys* (pp. 1–237). Routledge.
26. Mancini, F., Van Bruggen, A. H. C., Jiggins, J. L. S., Ambatipudi, A. C., & Murphy, H. (2005). Acute pesticide poisoning among female and male cotton growers in India. *International Journal of Occupational and Environmental Health, 11*(3), 221–232.
27. Zhang, Z., Huang, J., Yao, Y., Peters, G., Macdonald, B., La Rosa, A. D., Wang, Z., & Scherer, L. (2023). Environmental impacts of cotton and opportunities for improvement. *Nature Reviews Earth and Environment, 4*(10), 703–715.
28. Baskan-Bayrak, H., & Karakas, H. (2024). Morphology and chemical structure of a wool fiber. In S. Jose & G. B. Sabu Thomas (Eds.), *The wool handbook morphology, structure, properties, processing, and applications* (pp. 181–194). Woodhead Publishing.
29. Loh, J. W., Goh, X. Y., Nguyen, P. T. T., Thai, Q. B., Ong, Z. Y., & Duong, H. M. (2022). Advanced aerogels from wool waste fibers for oil spill cleaning applications. *Journal of Polymers and the Environment, 30*(2), 681–694.
30. Padhye, R., & Nayak, R. (2016). *Acoustic textiles*. Springer.
31. PETA. (2021). *Sheep farming and the wool industry's damaging environmental impact*. PETA.

32. Henry, B. K., Butler, W. D., & Wiedemann, S. (2015). Quantifying carbon sequestration on sheep grazing land in Australia for life cycle assessment studies. *The Rangeland Journal, 37*(4), 379–388.
33. Howden, S. M., White, D. H., & Bowman, P. J. (1996). Managing sheep grazing systems in southern Australia to minimise greenhouse gas emissions: Adaptation of an existing simulation model. *Ecological Modelling, 86*(2–3), 201–206.
34. Grishanov, S. (2011). Structure and properties of textile materials. In *Handbook of textile and industrial dyeing* (pp. 28–63). Elsevier.
35. Franck, R. R. (2001). *Silk, mohair, cashmere and other luxury fibres*. Elsevier.
36. Gupta, D., Agrawal, A., & Rangi, A. (2014). Extraction and characterization of silk sericin. *Indian Journal of Fibre and Textile Research, 39*(4), 364–372.
37. Tzenov, P., Cappellozza, S., & Saviane, A. (2022). Black, Caspian Seas and Central Asia Silk Association (BACSA) for the future of sericulture in Europe and central Asia. *Insects, 13*(1), 44.
38. Barcelos, S. M. B. D., Salvador, R., Guedes, G., Pinheiro, E., Moro Piekarski, C., & De Francisco, A. C. (2021). Socioeconomic and environmental aspects of the production of silk cocoons in the Brazilian sericulture. In *Sustainable fashion and textiles in Latin America*. Springer.
39. Nayak, R., Jajpura, L., & Khandual, A. (2022). *Traditional fibres for fashion and textiles: Associated problems and future sustainable fibres*. Elsevier.
40. Subramanian, K., Chopra, S. S., Cakin, E., Li, X., & Lin, C. S. K. (2020). Environmental life cycle assessment of textile bio-recycling—valorizing cotton-polyester textile waste to pet fiber and glucose syrup. *Resources, Conservation and Recycling, 161*(October 2019), 104989.
41. Lee, Y., Andrew Lin, K. Y., Kwon, E. E., & Lee, J. (2019). Renewable routes to monomeric precursors of nylon 66 and nylon 6 from food waste. *Journal of Cleaner Production, 227*, 624–633.
42. Duan, C., Wang, Z., Zhou, B., & Yao, X. (2024). Global Polyethylene Terephthalate (PET) plastic supply chain resource metabolism efficiency and carbon emissions co-reduction strategies. *Sustainability (Switzerland), 16*(10), 3926.
43. Thiemens, M. H., & Trogler, W. C. (1991). Nylon production: An unknown source of atmospheric nitrous oxide. *Science, 251*(4996), 932–934.
44. Emenike, E. C., Okorie, C. J., Ojeyemi, T., Egbemhenghe, A., Iwuozor, K. O., Saliu, O. D., Okoro, H. K., & Adeniyi, A. G. (2023). From oceans to dinner plates: the IMPACT of microplastics on human health. *Heliyon, 9*(10), e20440.
45. Chen, J. (2015). Synthetic textile fibers: regenerated cellulose fibers. In *Textiles and fashion materials, design and technology* (pp. 79–95). Woodhead Publishing.
46. Chang, S. J., Chen, C. J., Shih, T. S., Chou, T. C., & Sung, F. C. (2007). Risk for hypertension in workers exposed to carbon disulfide in the viscose rayon industry. *American Journal of Industrial Medicine, 50*(1), 22–27.
47. Shen, L., Worrell, E., & Patel, M. K. (2010). Environmental impact assessment of man-made cellulose fibres. *Resources, Conservation and Recycling, 55*(2), 260–274.
48. Drexler, H., Göen, T., Angerer, J., Abou-el-ela, S., & Lehnert, G. (1994). Carbon disulphide—I. External and internal exposure to carbon disulphide of workers in the viscose industry. *International Archives of Occupational and Environmental Health, 65*(6), 359–365.
49. Paul, D. R. (1968). A study of spinnability in the wet-spinning of acrylic fibers. *Journal of Applied Polymer Science, 12*, 2273–2298.
50. Skvortsov, I. Y., Chernikova, E. V., Kulichikhin, V. G., Varfolomeeva, L. A., Kuzin, M. S., Toms, R. V., & Prokopov, N. I. (2020). The effect of the synthetic procedure of acrylonitrile-acrylic acid copolymers on rheological properties of solutions and features of fiber spinning. *Materials, 13*(16), 3454.
51. Ve, İ., Larin, K. Ş., Kler, Ö. İ., Ne, İ. Ü. İ., & Tirma, B. (2024). *A research on yarn and fabric characteristics of acrylic/wool/angora blends*. AKRİLİK/YÜN/ANGORA LİFİKARIŞ IMLARINDAN ÜRET İLEN.

52. Yacout, D. M. M., Abd El-Kawi, M. A., & Hassouna, M. S. (2016). Cradle to gate environmental impact assessment of acrylic fiber manufacturing. *International Journal of Life Cycle Assessment, 21*(3), 326–336.
53. Kurian, J. V. (2005). A new polymer platform for the future—Sorona® from corn derived 1, 3-propanediol. *Journal of Polymers and the Environment, 13*(2), 159–167.
54. Tian, W., Huang, K., Zhu, C., Sun, Z., Shao, L., Hu, M., & Feng, X. (2022). Recent progress in biobased synthetic textile fibers. *Frontiers in Materials, 9*(December), 1–12.
55. Papageorgiou, G. Z. (2018). Thinking green: Sustainable polymers from renewable resources. *Polymers, 10*(9), 0952.
56. Hatti-Kaul, R., Nilsson, L. J., Zhang, B., Rehnberg, N., & Lundmark, S. (2020). Designing biobased recyclable polymers for plastics. *Trends in Biotechnology, 38*(1), 50–67.
57. Brody, H., & Division, I. C. I. F. (1976). The mechanical properties and structure of poly (m-methylene terephthalate). *Fibers, 14*, 263–274.
58. Farrington, D. W., Lunt, J., Davies, S., & Blackburn, R. S. (2008). Poly(lactic acid) fibers (PLA). In B. L. Deopura, R. Alagirusamy, M. Joshi, & B. Gupta (Eds.), *Textiles, polyesters and polyamides* (pp. 140–170). Woodhead Publishing.
59. Drumright, R. E., Gruber, P. R., & Henton, D. E. (2000). Polylactic acid technology. *Advanced Materials, 12*(23), 1841–1846.
60. Lunt, J., & Bone, J. (2000). Properties and dyeability of fibers and fabrics produced from polylactide (PLA)polymers. *AATCC Review, 1*(9), 20–23.
61. Kuciel, S., & Rusin-Żurek, K. (2024). Green Biobased Polyethylene Terephthalate (bioPET) composites reinforced with different lengths of basalt fiber for technical applications. *Fibers, 12*(9), 73.
62. Salvador, M., Abdulmutalib, U., Gonzalez, J., Kim, J., Smith, A. A., Faulon, J. L., Wei, R., Zimmermann, W., & Jimenez, J. I. (2019). Microbial genes for a circular and sustainable bio-PET economy. *Genes, 10*(5), 1–15.
63. Siracusa, V. (2020). Bio-Polyethylene (Bio-PE), Bio-Polypropylene (Bio-PP) and Bio-Poly(ethylene terephthalate) (Bio-PET): Recent developments in bio-based polymers analogous to petroleum-derived ones for packaging and engineering applications. *Polymers, 12*, 1641.
64. Sadeghi, B., Marfavi, Y., AliAkbari, R., Kowsari, E., Borbor Ajdari, F., & Ramakrishna, S. (2021). Recent studies on recycled PET fibers: Production and applications: A review. *Materials Circular Economy, 3*(1), 1–18.
65. Chinchillas-Chinchillas, M. J., Gaxiola, A., Alvarado-Beltrán, C. G., Orozco-Carmona, V. M., Pellegrini-Cervantes, M. J., Rodríguez-Rodríguez, M., & Castro-Beltrán, A. (2020). A new application of recycled-PET/PAN composite nanofibers to cement–based materials. *Journal of Cleaner Production, 252*, 119827.
66. Bataineh, K. M. (2020). Life-cycle assessment of recycling postconsumer high-density polyethylene and polyethylene terephthalate. *Hindawi Advances in Civil Engineering, 12*, 1–15.
67. Saleh, Y. (2016). Comparative life cycle assessment of beverages packages in Palestine. *Journal of Cleaner Production, 131*, 28–42.
68. Montava-Jorda, S., Lascano, D., Quiles-Carrillo, L., Montanes, N., Boronat, T., Martinez-Sanz, A. V., Ferrandiz-Bou, S., & Torres-Giner, S. (2020). Mechanical recycling of partially bio-based and recycled polyethylene terephthalate blends by reactive extrusion with poly(styrene-co-glycidyl methacrylate). *Polymers, 12*(1), 174. https://doi.org/10.3390/polym12010174
69. Parikh, D. V., Chen, Y., & Sun, L. (2006). Reducing automotive interior noise with natural fiber nonwoven floor covering systems. *Textile Research Journal, 76*(11), 813–820.
70. Nikolov, S., Lebensohn, R. A., & Raabe, D. (2006). Self-consistent modeling of large plastic deformation, texture and morphology evolution in semi-crystalline polymers. *Journal of the Mechanics and Physics of Solids, 54*(7), 1350–1375.
71. Vouyiouka, S. N., Karakatsani, E. K., & Papaspyrides, C. D. (2005). Solid state polymerization. *Progress in Polymer Science (Oxford), 30*(1), 10–37.

72. Atakan, R., Sezer, S., & Karakas, H. (2020). Development of nonwoven automotive carpets made of recycled PET fibers with improved abrasion resistance. *Journal of Industrial Textiles, 49*(7), 835–857.
73. Arafat, Y., & Uddin, A. J. (2022). Recycled fibers from pre- and post-consumer textile waste as blend constituents in manufacturing 100% cotton yarns in ring spinning: A sustainable and eco-friendly approach. *Heliyon, 8*(11), e11275.
74. Ütebay, B., Çelik, P., & Çay, A. (2019). Effects of cotton textile waste properties on recycled fibre quality. *Journal of Cleaner Production, 222*, 29–35.
75. Stanescu, M. D. (2021). State of the art of post-consumer textile waste upcycling to reach the zero waste milestone. *Environmental Science and Pollution Research, 28*(12), 14253–14270.
76. Gausa, M. (2024). *Knitting food: Food and eco-textiles new perspectives for sustainable urban*. Springer.
77. Elfaghi, A. M., Ilyas, R. A., & Sapuan, S. M. (2022). Corn: Its structure, polymer, fiber, composite, properties, and applications. *Polymers, 14*(20), 4396.
78. Jain, A., Rastogi, D., Chanana, B., Parmar, M. S., & Dhama, A. (2017). Extraction of cornhusk fibres for textile usages. *IOSR Journal of Polymer and Textile Engineering, 4*(1), 29–34.
79. Vijayan, V., Sathish, T., Saravanan, R., Kumar, K., Jadhav, G. K., Sharun, V., Elangovan, R., Chellamuthu, K., Singh, M., & Teklemariam, A. (2023). Waste coir nanofiller fused gallus-gallus fibres reinforced PMC. *Advances in Materials Science and Engineering, 2023*, 1–12.
80. Lingham-Soliar, T. (2017). Microstructural tissue-engineering in the rachis and barbs of bird feathers. *Scientific Reports, 7*, 1–13.
81. Parbhu, A. N., Bryson, W. G., & Lal, R. (1999). Disulfide bonds in the outer layer of keratin fibers confer higher mechanical rigidity: Correlative nano-indentation and elasticity measurement with an AFM. *Biochemistry, 38*(36), 11755–11761.
82. Mian, S. S. (2025). *Overview of lotus fiber: Types, sustainable properties and eco-friendly aspects* (June 2024).
83. Kalayci, E., Avinç, O. O., Bbozkurt, A., & Yavaş, A. (2016). Sustainable textile fibers obtained from agricultural wastes: Pineapple leaf fibers. *Sakarya University Journal of Science, 20*(2), 203–221.
84. Sachidhanandham, A. (2020). Textiles from orange peel waste. *Science and Technology Development Journal, 23*(2), 1730.
85. Abolore, R. S., Jaiswal, S., & Jaiswal, A. K. (2023). Green and sustainable pretreatment methods for cellulose extraction from lignocellulosic biomass and its applications: A review. *Carbohydrate Polymer Technologies and Applications, 7*, 100396.

Traditional Textile-Based Contemporary Fashion: A Roadmap to Cultural Sustainability

Yamini Jhanji

1 Introduction to Cultural Sustainability

Cultural sustainability is an umbrella term and can be considered as the fourth pillar of sustainable goals and principles with emphasis on:

- Cultural and artistic practices
- Knowledge dissemination and cultural heritage conservation
- Role of cultural traits and actions determining the sustainable livelihood of artisans

Cultural sustainability is based on the concept that usage and adaptation of cultural heritage by the current generation should be to the extent that allows understanding and preservation of embedded values and meanings associated with rich heritage by the upcoming generations. The central idea defining cultural sustainability is to never take more than necessary supporting the fact that overconsumption undoubtedly disrupts the natural balance, thereby leading to social and environmental catastrophic consequences.

Cultural sustainability in the fashion and interrelated textile field pertains to knowledge dissemination and rendering support pertaining to traditional know-how, competencies, skills, and legacy of fashion and textile cultures to the future generations. The harmony with the environment, seasons, and nature has been imbalanced owing to technologically driven processes and fast fashion. The textile and fashion industry in recent times is gradually transitioning toward slow fashion advocated by indigenous cultures thus taking a leap toward cultural sustainable aspects.

Apart from culturally sustainable aspect, indigenous craftsmanship can strike a chord among consumers of varied socio-economic statures since artisan fashion is

Y. Jhanji (✉)
Department of Textile Technology, TIT&S, Bhiwani, Haryana, India

S. S. Muthu (ed.), *Cultural Sustainability in Fashion*, Sustainable Textiles: Production, Processing, Manufacturing & Chemistry,
https://doi.org/10.1007/978-3-032-03207-2_6

at par with luxury fashion. Artisan fashion can be regarded as a form of luxury owing to remarkable similarity in terms of lead time, necessary knowledge and training of craftsmen involved, complexity of manual work, raw material usage, value addition, and appreciation of end products.

2 Cultural Appropriation and Issues Related to Artisans' Well-Being

Cultural appropriation is associated with adoption of elements of another culture neither acknowledging the actual creators nor understanding or respecting the significance and cultural values inherent to a particular craft or tradition. A key factor depriving the artisans and craftsmen of their due credit and economic empowerment is cultural appropriation.

Recent times have witnessed upsurge in design collections that are projected as designer's novel work but are actually knock offs of artisans' workmanship obscuring the true cultural context. The misrepresentation of the cultures involved, via adoption of elements and practices from minority cultures and communities by design houses and labels need to be stringently addressed to avert commercialization and distortion of cultural symbols.

Apart from threats posed to novelty and creativity of artisans by cultural appropriation, the artisans and skilled workforce is exposed to several other challenges (as enlisted below) making their quest for survival and livelihood gruesome.

- Globalization and mass production
- Capital and time-intensive handcrafted processes
- Inability to adapt to changing market trends
- Environmental constraints and limited access to raw materials
- Reluctance of next generations to learn and pursue the conventional practices
- Ignorance about advanced technological and digital processes redefining the textile and fashion world

The humongous technological and digital advancements have left the craftspeople in a state of predicament regarding integration of novel and advanced processes without compromising the pristine attributes of their craftsmanship. Simultaneously, this paradigm shift should not be perceived as passive adoption of specific technologies, rather it should be relegated as a collaborative path of re-engineering, idea and technology transfer, and knowledge by:

- Delineation of similarity and contrast
- Focus on preservation and innovation
- Creating based on typicality and novelty

Cultural sustainability in fashion requires taking into account cultural appropriation and associated issues of ownership, authorship, threatening of cultures, and their material representations. Cultural appropriation leads to commercial exploitation of traditional craft with artisans not given due credit for their craftsmanship and thus devastating social and economic implications for indigenous communities.

Cultural appropriation can be circumvented by meaningful dialogue and collaboration of designers with indigenous craftsmen. Collaboration is a valued aspect in textile and fashion domain that inculcates creativity and novelty. However, sensitivity, due acclamation and mutual respect is of utmost importance as designers collaborate with artisans. The integrity and significance of cultural expressions within the textile and fashion industry can be maintained by mindfulness of designers and consumers of the cultural origins of designs they create and embrace.

Thus the need of the hour is to empower artisans and craftsmen by uplifting their social and economic stature, fostering self-reliance, encouraging them to pass their rich legacy to future generations, and providing a holistic showcasing platform for their fine workmanship. Undoubtedly, apart from designers and manufacturers, the consumers can also play their part in supporting the indigenous artisans and craftsmen by giving precedence to home grown, indigenous, and local artisanal brands and design collections over luxury and high-end international brands.

Cultural sustainability can thus be imbibed and practically implemented when consumers develop pungent for domestically developed artisanal ensembles and are wary to make deliberate purchases to support local artisans and craftsmen. Consequently, Indian craftsmen hold a fair chance not to earn a means of livelihood without thinking of alternate career avenues but also to showcase their skills, artforms thereby preventing the gradual dying out of the artforms connected to our roots.

3 Traditional Textiles: Roadmap for Cultural Sustainability

Traditional textile craftsmanship can serve as a roadmap for culturally sustainable design collections. Adoption of cultural sustainable principles via resurgence of indigenous textiles is instrumental in safeguarding the environment and natural resources of the region along with enhancing the livelihoods of the local community.

Moreover, the magnificence and aura of traditional textiles has mesmerized consumers, textile manufacturers, and designers since time immemorial. The traditional textiles play a pivotal role in the country's economic growth and can help in earning export-based revenue owing to the humongous export avenues in global arena. The worldwide clamor and preference for indigenous crafts and traditional textiles has fostered government bodies, textile designers, manufacturers, and other stakeholders to endorse the heritage of traditional textiles globally. Apart from economic considerations and avenues of earning foreign exchange through export of indigenous textiles and handicrafts, the ardent need to adopt sustainable tools and principles in textile and fashion supply chain foster the designers, manufacturers, and other stake holders to look for ecofriendly options substituting the conventional, staple raw materials that are deleterious to environment. Additionally, the lack of proper strategies for recycling and upcycling of discarded clothing and reckless, mindless purchasing, usage, and disposal by consumers leads to the clothing ending as landfills thereby contributing to carbon emissions and adverse environmental impact. It is thus imperative to adopt and follow sustainable principles in the textile and fashion supply chain [1, 2]. Consequently, majority of the textile units and fashion brands have been prompt in transitioning toward sustainable principles. The first pathway for meeting sustainable goals is at the raw material stage and thus natural and synthetic fibers that demand wasteful utilization of natural resources are being replaced by sustainable alternatives like soya bean, pineapple, hemp, jute, bamboo, recycled polyester, and organic cotton to name a few [3–6]. Likewise, a gamut of fashion designers and brands have been trailblazers in creating apparel and accessory line based on recycling, reuse, and upcycling principles. Similarly, novice as well as amateur designers attempt to promote traditional textiles, revive some traditional attires that lost their sheen in the wake of modernization and automation and work in collaboration with artisans (Fig. 1). Designers have been proactive in supporting artisans and craftsmen by associating with them for their runway collections showcasing foray of traditional and handcrafted ensembles.

Designers have been blending traditional artforms and techniques with modern sensibilities for creation of exquisite contemporary wear. Furthermore, the driving force for designers to endorse traditional textiles is also rooted in the inherent sustainable nature of Indian sartorial heritage such as weaving, hand embroidery and painting, textile printing, and dyeing or embroidery. The traditional crafts are synonyms with the emotional and cultural sensibilities of artisans that extend beyond

Fig. 1 Majestic usage of traditional textiles and motifs on runways inspired by artisans' fine workmanship

the rat race of tangible product development. Artisans give precedence to natural dyestuff, colors, and hand-spun textiles for most of the craft activities further strengthening the cultural sustainable aspects. Moreover, traditional attires such as dhotis and sarees follow a zero-waste pattern making principle by being designed for longevity, easy drapability, and fit irrespective of changing body physiology.

Accordingly, the combined efforts and association of designers and budding entrepreneurs with artisans have turned the spotlight on rich Indian heritage, art, tradition, and textiles reviving the techniques in their recurring styles as well. Additionally, a sustained source of employment and livelihood to artisans along with opportunity to showcase their unparalleled skills internationally is ensued by lucrative collaboration.

High-end designers like Manish Malhotra, Anita Dongre, Rahul Mishra and Anavila Misra, Rina Singh, and Gautam Gupta have been instrumental in working hand in hand with artisans thereby launching exclusive design collections that speak volumes about the rich heritage of traditional textiles and hand-curated workmanship [7, 8].

The collaboration of designers with artisans in developing design collections narrating the story of rich heritage of traditional textiles is undoubtedly a synergistic strategy for both the stakeholders. As such the traditional textile inspired designs boast of sustainability and slow fashion since artisans working with traditional textiles prefer virgin, indigenous materials and methods unlike the factory-made products [8–11].

Integration of traditional textile know-how into contemporary fashion supply chains and fostering meaningful association with artisans and local communities can go a long way in eliminating fast fashion, obsolescence, consumerism, misappropriation, and unethical exploitation of resources and workforce deep rooted in fashion world.

4 Fashion and Craft Synergy

Creation of culturally sustainable fashion in recent times demand integration of traditional knowledge and practices with contemporary fashion and textile supply chains so as to create ample opportunities for upcoming generations. Moreover, the deleterious impact of mainstream fashion can be averted and well-being of local communities can be ensued through synergies between fashion and craft.

The synergistic relationship between fashion and craft is by Cultural Intellectual Property Rights Initiative® by coining the 3Cs Rule:

- Consent pertains to prior, free, and informed consent of artisans, craftsperson, or local communities.
- Credit is related to due acknowledgement for design inspiration from parent source, i.e., indigenous workforce, craftsmen, artisans, etc.

- Compensation relates to monetary or nonmonetary remuneration as a token of appreciation and acknowledgement for artisans.

Cultural sustainability can be thus successfully implemented if fashion industry embraces the novelty, power, and manual artistry. This in turn can be accomplished by promulgating 3C rule for designer- artisan collaborations thereby safeguarding respect and dignity of craftsmen along with ethical design creations that support cultural continuity.

Also, preservation, endorsement, and safeguarding cultural heritage in the form of traditional textiles ought to be a collective endeavor of designers, manufacturers, governments, and non-profit organizations. Additionally, consumers can serve as stewards in upholding cultural sustainable practices by making informed choices, advocating fair trade practices, and actively connecting and economically empowering the artisans. The mindfulness and engagement of consumers in present times can lay a strong foundation for endurance of craftsmanship, expertise, and narratives associated with traditional textiles across generations [12, 13].

5 Cultural Sustainable Avenues via Collaborative Initiatives of Designers and Artisans

The current fashion landscape showcases deep imprints of culture on fashion with a myriad of top-notch fashion designers working toward revival of traditional textiles and crafts. The vibrant and intricate Indian sarees boasting of nation's diverse regional customs and rich cultural heritage have been glorified on various national and international platforms including Olympics and Cannes apart from runways. The distinct designs, colors, and styles of sarees prompt designers to work in unison with weavers and artisans to integrate tradition with contemporary fashion thereby preserving cultural legacies.

The impact of culture on fashion design is further reflected in the choice of raw materials in the clothing line designers create. The incorporation of materials from various cultures featuring their unique traditions and techniques can be a roadmap for designers to create distinctive, novel design collections. Traditional textiles such as cotton and silk are abundantly explored for Indian lifestyle accessories. Likewise, handloom fabrics interweaving history, tradition, and artistry is extensively explored for designing couture as well as casual designs. Moreover, weaving is an age-old tradition practiced as a means of livelihood across generations of various communities in different nooks and corners of the country. Handloom weaving and craft practices by local artisan communities boast of strong cultural identity and are sustainable too since the manually operated looms unlike the power looms do not rely on power or fossil fuels and thus result in low carbon footprint.

A gamut of fashion designers and brands have been trailblazers in promoting traditional textiles, reviving some traditional attires which lost their sheen in the wake of modernization and automation and work in collaboration with artisans. The collaboration of designers with artisans in developing design collections narrating the story of rich heritage of traditional textiles is undoubtedly a synergistic strategy for both the stakeholders. As such the traditional textile inspired designs boast of sustainability since artisans working with traditional textiles prefer virgin, indigenous materials and methods unlike the factory-made products. The clothing brands and designers in an attempt to adopt sustainable principles and design ethically are prompt in brand positioning and lucrative collaborations. Additionally, the designers are coming up with the innovative concept of trans-seasonal capsule collections that flaunt of multicultural Indian brigade celebrating the prowess and affluence of Indian craftsmanship. Also, the capsule collections are designed with the central idea of encouraging the consumers to shop mindfully and mix and match their ensembles. H& M collaborated with Sabyasachi launching travel themed capsule collection offering a grandeur of bohemian era featuring warm colors, shimmering, golden embroidered fabrics and flowing silhouettes (Figs. 2 and 3). Rahul Mishra, an ace ethnic wear designer considers his artisans like butterflies and believes in bringing out the best potential of artisans by continued motivation thereby enabling the karigars to preserve their craftsmanship without displacing them from their native hometowns.

Fig. 2 Online platforms for traditional textiles

Fig. 3 Lucrative association of designers & artisans

Anita Dongre, the face behind the popular sustainable brand "GrassRoots" is a staunch supporter and promoter of indigenous textile craftsmanship by artisans. The designer believes in design intervention and is proactive in rendering assistance to artisans in applying their talents in diversified contemporary collections. Joining hands with artisans, Dongre has been successfully catering to modern, global shopper via recreation of design and color combinations from traditional Indian attires.

Rohit Bal launched his Ghevar collection for Biba that speaks volumes about designer's passion to ooze the grandeur and royalty of Indian heritage portraying signature lotus and peacock motifs, rich fabrics, flattering silhouettes and classic colors.

The designers at Pero, the upcycled brand ardently supports localization of trends thereby using the fabrics and transforming them to chic, runway ready attires with inputs of local artisans. The age-old traditions of handmade crafts practiced by artisans are woven into intricate designs completed with hand embroidery [14–16].

Ritu Kumar, an ace Indian designer with affirmation to revive traditional textiles and foster its acceptability among young consumers prudently design the clothing line to narrate the richness of traditional textiles along with modern twist by choosing unconventional silhouettes. The contemporary clothing line Kumar designed combined the charm of international silhouette and Indian textiles. A perfect amalgamation of old-world and modern-day influences was reflected by mindful utilization of color palette of earthy tones such as burgundy and sap greens, and fabrics like woven jacquards, cotton, chanderi, and silk treated with traditional vegetable dyes.

Gaurang Shah infused life in the paintings of an ace painter Raja Ravi Varma by transferring mythological scenes from his paintings onto hand spun khadi saree pallus. Shah showcased the exhibits in "Santati: Khadi—A Canvas" exhibition featuring khadi as a wearable art with interpretation of 30 paintings of Raja Ravi Varma's work utilizing natural dyes for replicating the paintings in khadi and recreating a similar color palette.

Shruti Sancheti, an Indian ethnic wear designer, in an attempt to revive traditional Indian weaves has come up with a pret-a-porter label named Pinnacle. The label flaunts of ethnic yet modern and edgy ensembles that features vibrant colors and generous use of natural fiber-based fabrics and traditional dyeing and printing techniques [2, 3].

Injiri, the brand by Chinar Farooqui, is a tribute to India's rich textile heritage primarily focusing on organic materials and natural dyes. The brand flourishes in the current global market due to masterful blending of textile traditions with contemporary design sensibilities.

Designers at Fabindia and Good Earth are partnering with artisans to introduce and endorse clothing and accessory line exclusively made through traditional techniques. Several homegrown small- and medium-scale businesses are also playing a pivotal role in promotion of traditional art by their meaningful collaborations with the skilled artisans. The Indian Ethnic Co is a mother-daughter brand who aspire to make Indian fashion responsible, sustainable, and truly handcrafted, started designing apparels out of their love for handloom fabrics. However, their relentless efforts brought the business to such magnitude that they are supporting livelihoods of nearly 1000 artisans and processing 3000 monthly orders. Okhai, a brand known for its vibrant and elegant range of traditional textile based contemporary and fusion wear was launched by Kirti Poonia with a sole objective to support rural women artisans. The company has witnessed growth and scaling ever since its inception claiming to be supporting and empowering 24,000 women artisans.

Tjori, a brand synonymous with elegant hand-curated traditional textiles was launched by Mansi Gupta, with an aim to cater to the global demand for Indian handcrafted products. Gupta researched to conclude that there was growing demand for Indian handicrafts abroad with no Indian brand fulfilling the consumer aspirations and driving Indian audience. The lack of knowledge about the traditional and heritage handicrafts of India on a global platform paved the way for brand to flourish globally thus serving as bread earner for more than 500 artisans and craftsmen.

Manchaha, an initiative by Jaipur Rugs strives the revival of the age-old tradition of hand-knotted carpets by collaborating with local weavers. The global promotion of hand-spun rugs by the designer intervention fosters the weavers to renew the passion for weaving [17].

Textile revivalists Swati Agarwal and Sunaina Jalan have been working with weavers to weave Gyaser fabric suitable for draping as sari. The silk-brocade fabric featuring narrow width and high thickness served as popular embellishment in monasteries unlike the transformation rendered by weavers' intervention to create the fabric in varied thickness and width intended for saree.

Another lucrative co-creation framework aimed at stimulating artisan-designer collaborations is WhyWeCraft campaign under the flagship of Cultural Intellectual Property rights Initiatives (CIPRI). The merger of contemporary design thinking with traditional textile knowledge and techniques through the campaign affirms the cultural sustainable practices adopted by European textile and fashion market. It can thus be recapitulated that designers can truly be espousal of cultural sustainability by due consideration of cultural significance in each aspect of their design collections right from raw material procurement to prudent selection of patterns, values, and symbols. The steward approach of designers can ensure design collection that apart from being functional and stylish adheres to principles defining cultural sustainability [18, 19].

6 Government's Role in Promotion of Traditional Textiles: A Roadmap to Cultural Sustainability

Undoubtedly, the designers are upbeat in associating with local artisans to promote their indigenous crafts and offer them better exposure, wages, and a sustained source of livelihood. Providentially, Ministry of Textiles has also emerged as an ardent promoter of Indian Traditional Textiles and indigenous crafts. The classic example of endorsing the traditional textiles was the Republic Day tableau that showcased the evolution of traditional textiles to futuristic technical textiles. The tableau was segregated in front and rear with former featuring shuttle, charkhas, cotton balls, and silk cocoon on top while the rear part featured colored yarns being converted into fabric by looms and in turn into different traditional and technical apparels. A weaver weaving yarn dyed fabric in the handloom was another highlight of the tableau. Additionally, the government in the pursuit to entice teenagers and millennials to embrace handloom and traditional textiles came up with an interesting campaign of Selfie with Handloom thereby encouraging them to dress up in

Fig. 4 Government's initiatives to promote traditional and hand-curated textiles

traditional textile-based attire and post their selfies on their social media platforms (Fig. 4). Likewise, the then Minister of Textile Ms. Smriti Irani collaborated with women wear ethnic brand "Biba" to popularize khadi and handloom textiles. A step ahead toward massive endorsement of traditional textiles was putting up virtual Bharat Parv featuring the rich heritage of traditional textiles of different Indian states and regions.

Furthermore, the Government of India taking into consideration the employment potential and profitability of textile sector is prompt in launching a myriad of lucrative schemes such as Skill India, Make in India, Samarth for massive promotion, investment and creation of jobs, business, and skill upscaling in textile sector. Additionally, the National Technical Textile Mission (NTTM) launched in 2020 under the flagship of Ministry of Textiles is yet another government initiative to promote higher education, ground-breaking research, and innovation in textiles and to build expertise in technical textiles. A total of approximately 74 research proposals related to specialty fibers and technical textiles valuing Rs 232 Cr have been approved under NTTM with an objective of positioning India as key player in technical textiles across the globe [20, 21].

7 Role of Undergraduate and Textile Enthusiasts in Promotion of Traditional Textiles

Academicians and researchers can also contribute to the promotion of traditional textiles by initiatives such as:

- Identification of artisan clusters serving in nearby semi-urban and rural regions.
- Organizing regular cluster visits, workshops, and awareness programs for artisans.
- Equipping artisans with latest technologies and modernizations in textile domain.
- Collaborating with artisans for undergraduate and postgraduate projects and dissertations.

With these objectives in view, textile and fashion undergraduate students at TIT&S, Bhiwani, were mentored to identify the indigenous textile crafts and related activities practiced by artisans in and around Bhiwani district. The students collaborated with artisans and created contemporary wear embellished with varying surface ornamentation techniques. The exercise fostered zeal among artisans to exhibit their creativity in a transformed contemporary clothing line and that too on a bigger platform.

Accordingly, the students conducted unstructured and semi-structured interviews with artisans to keep themselves abreast of artisans' challenges and expectations. Most of the artisans were unaware of effective strategies to promote their work on online platforms. Likewise, consumer and market scans were conducted to understand consumers' acceptance and spending capacity for traditional textile-based clothing line. The target consumers for design collection were female professionals who had penchant for authentic handicrafts and traditional textiles. A myriad of eco-friendly, sustainable apparels and accessories were developed from left over fabrics from artisans (Fig. 5). Furthermore, artisans were encouraged to create hand-curated complimentary jute-based jewelry line. The design collection was affordable with minimal selling prices decided for each entity owing to the maximum utilization of upcycled and recycled raw materials in the same. The subjective evaluation of the end-products implied that fashion conscious professional females have predilection for minimalistic fashion preferring minimal embellishments and traditional motifs like madhubani and shibori prints in their attires. The design exercise served as best of both worlds not just by offering quirky yet elegant traditional textile-based design collection for female consumers but also enabling artisans to refine their creative endeavors by amalgamating age-old traditional techniques with a tinge of modernization.

Fig. 5 Hand-crafted sustainable end-products in collaboration with artisans

7.1 Design and Development of Traditional Design Collection: Desi Treasure

Traditional textile-based design collection "Desi Treasure" comprising of hand-curated home decor, apparels, and accessories was developed in collaboration with local artisans. The surface ornamentation techniques chosen for the ensembles included hand embroidery, applique work, patch work, tie and dyeing, and hand painting. The inspiration for the design conceptualization emerged from brainstorming sessions with artisans, local dyers, and indigenous craftsmen. Extensive utilization of vibrant colors, simple silhouettes, and nature-inspired motifs was a deliberate attempt to strike a chord with millennials and advocate the concept of cultural sustainability. Figure 6 shows the traditional textile-based, sustainable design collection (home decor, apparels, and accessories) utilizing the varied surface ornamentation techniques.The inspiration for embroidery and hand painting motifs was garnered from artisans and indigenous craftsmen who pursue this traditional artform as means of their livelihood.

The respondents were asked to rate the designed products on the basis of their functionality and aesthetic appeal. Subjective evaluation of the designed ensembles suggested that most of the respondents chose hand embroidery as the preferential embellishment technique over other techniques chosen for the study.

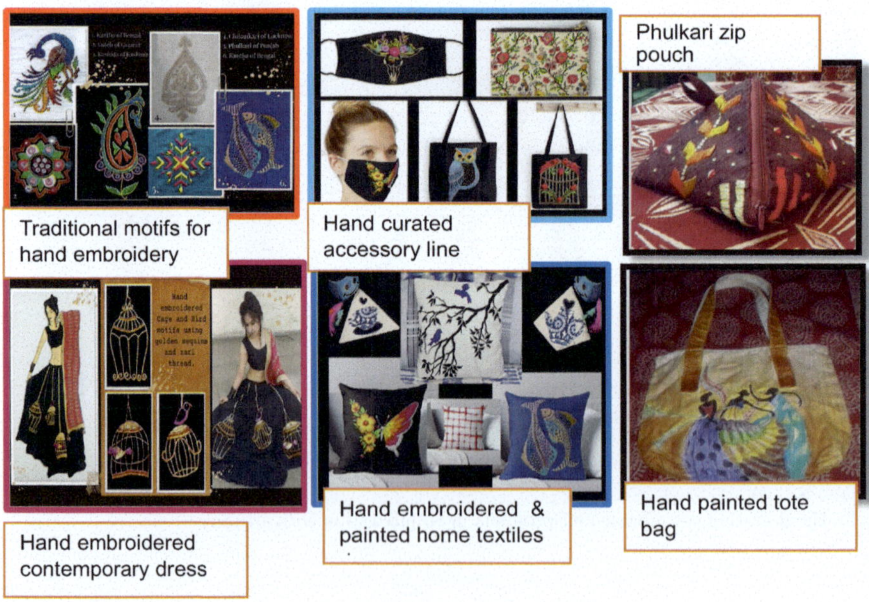

Fig. 6 Hand-curated traditional textile-based, sustainable design collection

8 Conclusion

The upsurge of experimental and unconventional fashion among consumers prompts the designers and other stakeholders associated with fashion world to prioritize likes and aspirations of fashion connoisseurs. Also, today's consumers wish to stay connected to their roots and prefer pristine crafts and artforms. The rich heritage of traditional textiles flaunts the amalgamation of incredible and exclusive skill, heritage, and dynamism via meaningful association and inclusion of indigenous artisans and craftsmen in the mainstream.

Cultural sustainability can be imbibed and practically implemented when consumers develop pungent for domestically developed artisanal ensembles and are wary to make deliberate purchases to support local artisans and craftsmen. Consequently, the Indian craftsmen hold a fair chance not only to earn a means of livelihood without thinking of alternate career avenues but also to showcase their skills and artforms thereby preventing the gradual dyeing out of the artforms connected to our roots.

The ardent need of the hour is thus endorsement of indigenous craftsmanship and traditional textiles by re-creation of the techniques with modern/contemporary design aesthetics to support the artisans and workforce financially and providing a larger audience and global markets for their hand-curated end-products.

References

1. Jhanji, Y. (2021). Design & development of traditional & sustainable contemporary women wear. *Latest Trends in Textile and Fashion Designing, 4*(2), 000183. https://doi.org/10.32474/LTTFD.2021.04.000183
2. Jhanji, Y., & Hooda, A. (2022). Rebooting post pandemic fashion by endorsing vocal for local. *Latest Trends in Textile and Fashion Designing, 4*(4), 000193. https://doi.org/10.32474/LTTFD.2022.04.000193
3. Jhanji, Y., & Kaur, G. (2022). Design & development of ornate upcycled fashion ensembles. *Latest Trends in Textile and Fashion Designing, 4*(4), 000194. https://doi.org/10.32474/LTTFD.2022.04.000194
4. Jhanji, Y., & Grewal, D. (2020). *Revival of madhubani painted textiles*. https://www.fibre2fashion.com/industry-article/8682
5. https://b1-akt.com/FR/wp-content/uploads/2019/08/Cultural-Sustainability-in-Fashion-Workshop.pdf
6. *Promoting sustainability in traditional textiles through consumer engagement*. (2024). https://www.iiad.edu.in/
7. https://curiositysavestheplanet.com/sustainable-textile-and-clothing-practices-from-traditional-indian-culture/
8. Pradhan, S., & Khandual, A. (2020). Community, local practices and cultural sustainability: A case study of sambalpuri ikat handloom. In *Sustainability in the textile and apparel industries* (pp. 121–139). Springer.
9. Vacca, F., Colombi, C., D'Itria, E., et al. (2019). *Education and research: Benchmarking report*. https://www.artun.ee/app/uploads/2020/02/IO1_BENCHMARKING-REPORT_-Fashion-SEEDS.pdf

10. https://yourstory.com/smbstory/indian-brands-artisans-traditional-handicraft-products
11. https://www.architecturaldigest.in/content/independence-day-special-9-brands-designers-indian-arts-crafts-make-us-proud
12. https://textilevaluechain.in/news-insights/apparel-fashion-retail-
13. Ketki. (2021). *Sustainable textile and clothing practices from traditional Indian culture*. https://textilevaluechain.in/news-insights/apparel-fashion-retail-news/fashionvaluechain/reviving-traditional-textiles-sustainable-practices-and-future-potential
14. Shah, D. (2024). *Cultural integrity in fashion: Embracing heritage and sustainability in design*. https://www.fibre2fashion.com/industry-article/9895/cultural-integrity-in-fashion-embracing-heritage-and-sustainability-in-design
15. Brown, S., & Vacca, F. (2022). Cultural sustainability in fashion: reflections on craft and sustainable development models. *Sustainability: Science, Practice and Policy, 18*(1), 590–600. https://doi.org/10.1080/15487733.2022.2100102
16. https://made-in.org/craft-thinking/cultural-sustainability
17. https://ourculturemag.com/2025/02/27/the-evolving-intersection-of-art-fashion-and-sustainability-in-contemporary-culture/
18. https://www.culturalsustainabilityinfashion.info/
19. https://www.technopak.com/rising-stars-indian-fashion-designers-making-global-impact-in-2024/
20. https://www.outlooktraveller.com/experiences/heritage/reviving-tradition-how-indian-designers-are-saving-indigenous-textiles-from-fast-fashion
21. https://www.designpataki.com/artisanal-brands-taking-indian-textiles-world/

Index

S. S. Muthu (ed.), *Cultural Sustainability in Fashion*, Sustainable Textiles: Production, Processing, Manufacturing & Chemistry,
https://doi.org/10.1007/978-3-032-03207-2

The manufacturer's authorised representative in the EU is Springer Nature Customer Service Centre GmbH, Europaplatz 3, 69115 Heidelberg, Germany. If you have any concerns regarding our products, please contact ProductSafety@springernature.com

Printed and bound by CPI Group (UK) Ltd, Croydon, CR0 4YY
16/09/2026
02217816-0001